W9-BXH-279

SAFE DRINKING WATER ACT

Amendments, Regulations and Standards

Edited by

Edward J. Calabrese
Charles E. Gilbert
Harris Pastides

Library of Congress Cataloging-in-Publication Data

Safe Drinking Water Act.

Bibliography: p.
Includes index.
1. Drinking water--Law and legislation--United States. I. Calabrese, Edward J.
II. Gilbert, Charles E. III. Pastides, Harris.
KF3794.S24 1989 346.7304'69122 88-26679
ISBN 0-87371-138-6 347.306469122

Third Printing 1990

Second Printing 1989

LEWIS PUBLISHERS, INC.
121 South Main Street, Chelsea, Michigan 48118

PRINTED IN THE UNITED STATES OF AMERICA

To Mr. and Mrs. Ray Varanka
for their constant encouragement and support.

To Sharon and Sarah Gilbert
with great affection.

To Patricia, Katharine, Andrew
for their support.

Edward J. Calabrese is a board-certified toxicologist who is professor of toxicology at the University of Massachusetts School of Health Sciences, Amherst. Dr. Calabrese has researched extensively in the area of host factors affecting susceptibility to pollutants. He is the author of more than 240 papers in referred journals and 10 books, including *Principles of Animal Extrapolation, Nutrition and Environmental Health Vols. I and II, Ecogenetics,* and others. He has been a member of the U.S. National Academy of Sciences and NATO Countries Safe Drinking Water committees and, most recently, has been appointed to the Board of Scientific Counselors for the Agency for Toxic Substances and Disease Registry (ATSDR).

Dr. Calabrese was instrumental in the conceptualization and development of the Northeast Regional Environmental Public Health Center and was appointed its first director. The center's mission includes communication, education, and research.

Charles E. Gilbert is a research associate in toxicology at the University of Massachusetts School of Health Sciences, Amherst. He received his BS and MSc and is a candidate for a PhD from the University of Massachusetts.

He was the assistant director for the Childhood Lead Poisoning Prevention Program, Massachusetts Department of Public Health, where he directed improvements in environmental management, case management, and education programs. His research interests are in the area of factors that affect human susceptibility to biological, chemical, and physical agents and how these affect health.

Charles Gilbert worked in the development of the Northeast Regional Environmental Public Health Center and was appointed its first assistant director. This center is a cooperative organization of the New England Public Health Departments and the University of Massachusetts School of Health Sciences.

Harris Pastides is an associate professor of epidemiology at the University of Massachusetts School of Health Sciences, Amherst. He was awarded a BS degree in biological sciences from the State University of New York at Albany, and MPH, MPhil, and PhD degrees in epidemiology from Yale University.

Dr. Pastides has been on the faculty of the University of Massachusetts since 1980 and associate director of the Northeast Regional Environmental Public Health Center since its inception in 1985. He has conducted extensive research in a large number of public health fields, including cancer epidemiology, musculoskeletal epidemiology, and occupational health. He is an elected member of the American College of Epidemiology and a regular consultant to the National Academy of Sciences Board on Science and Technology for International Development, as well as to several state health departments and to industry.

Preface

The quality of drinking water continues to be a major public health concern. The public now recognizes that drinking water quality can be degraded by numerous processes, such as the effluent from leaking underground storage tanks, agricultural runoff, improper industrial practices, corrosive plumbing, and the drinking water treatment process itself. In addition, it is now known that naturally occurring levels of radon, arsenic, fluoride, and other agents may present significant public health concerns.

Congress responded to these concerns by passing the Safe Drinking Water Act (SDWA) in 1974 and the Safe Drinking Water Act Amendments of 1986, both of which give testimony to the sustained concern and involvement of Congress in ensuring that safe drinking water be a national priority.

By communicating how the EPA has responded to the mandates of Congress (mainly the SDWA Amendments of 1986) this book approaches the topic of safe drinking water differently from other books describing the presence of pollutants in treated drinking water. Moreover, while this book focuses on how EPA has responded to the national concerns over drinking water quality and congressional action, it also includes an independent, non-EPA viewpoint, with particular attention to public health concerns associated with drinking water treatment practices, especially disinfection (Chapter 12). Specifically, this book includes:

1. how EPA defines the magnitude of public health concerns and drinking water quality
2. what EPA believes are the public health priorities for drinking water exposures in the United States
3. how these priorities are translated into programs
4. the likely benefits and costs of these programs
5. the time frame within which these goals will be achieved

Chapter 1 summarizes what is and will be involved in achieving safe drinking water. Chapter 2 describes the historical development of drinking water regulations through the present. Chapter 3 summarizes the directives of the Safe Drinking Water Act Amendments of 1986, describes the philosophy of and some of the methods used by the EPA in developing the drinking water regulations, and defines some of the terms used in the regulations.

Chapters 4 through 9 discuss each phase of the regulatory program in turn. Specific problems associated with volatile organic chemicals, synthetic organics, inorganic chemicals, and microbiological contaminants are assessed in Chapters 4 and 5. The unique characteristics of radionuclides and their regulation are treated in Chapter 6. The disinfection process and its resultant disinfection by-products are presented in Chapter 7. The contaminant selection process and the additional contaminants to be regulated by 1989 and 1991 and in future years are discussed in Chapters 8 and 9.

EPA's Office of Drinking Water's Health Advisory Program is explained in Chapter 10. The record of public water system compliance with the primary drinking water regulations is detailed in Chapter 11. Chapter 12 offers a nongovernmental perspective on the general quality of drinking water in the United States and how this is affected by a wide range of drinking water treatment technologies, with particular attention to the multifaceted public health challenges associated with disinfectant technologies and their possible by-products. For ready reference to the reader, we have also included as Appendix 1 a copy of the Safe Drinking Water Act as amended by the Safe Drinking Water Act Amendments of 1986, adapted from a publication prepared by Camp Dresser & McKee Inc.

Edward J. Calabrese
Charles E. Gilbert
Harris Pastides

Amherst, Massachusetts
March 1988

Acknowledgments

We would like to acknowledge the overall efforts of the United States Environmental Protection Agency, Office of Drinking Water Criteria and Standards Division, Washington, DC, and the Region I Drinking Water Office, Boston, Massachusetts. We also acknowledge Carol Cady for her work in the compilation of this book. Finally we thank Paula Goodhind for coordinating the processing of the manuscripts and communication with the authors, and Linda Curtis for word processing.

Contents

SAFE DRINKING WATER ACT

Amendments, Regulations and Standards

CHAPTER 1

Achieving Safe Drinking Water: Summary and Recommendations

Bailus Walker

The first drinking water standard, designed to protect the public against acute bacterial diseases, was set in 1914 by the U.S. Public Health Service. Over the years, it has been revised to include source protection, chemical standards, and a radioactivity standard. In 1974, Congress passed the Safe Drinking Water Act, which was designed to protect groundwater from contamination by organic and inorganic chemicals, radionuclides, and microorganisms. In 1986, the act was amended. Support for such legislation has been high, since nearly half of the nation's population relies on groundwater as a primary source of drinking water. Our national effort these past two decades to improve water quality continues to show results; however, new challenges remain.

Few areas of the country are without water quality problems. Because contaminated groundwater is often located near industrialized, populated areas, exposure is more likely. More and more point sources are under control, but nonpoint sources remain a threat to groundwater. Potential sources of contamination include agricultural runoff (pesticides and fertilizers), landfills, underground storage tanks, and septic tanks.

Although groundwater contamination is extensive and its potential for limiting growth is significant, it has been difficult to assess, largely because of inadequate data. Due in part to the efforts of the U.S. Environmental Protection Agency, inventorying and monitoring of certain sources of contamination is improving. Monitoring will be increasingly important as we gauge the extent of groundwater contamination.

Regulation of chemical carcinogens in groundwater supplies is the subject of much debate. Carcinogen regulations must reflect our desire to reduce the incidence of cancer and our uncertainty about causation, particularly in cases of low-dose levels of suspected carcinogens in water supplies.

Our agenda for a strong Safe Drinking Water Act must include several components:

1. We must attempt to monitor and regulate the broadest range of compounds possible, recognizing the great diversity of contaminants that pose a threat to our groundwater. At present, many substances known to occur in groundwater are not regulated.

2. We need to develop a comprehensive information management system to help us evaluate the vast quantities of data that have been generated. Such a system can be invaluable in helping to detect trends and to take preventive measures where necessary. To be effective, however, such an information system must take into account that pollutants released into the environment are distributed throughout the environment. Pollutants in the air or in soil often end up in the groundwater.

3. States need to adopt comprehensive state groundwater plans and regulations. These plans need to make the protection of groundwater quality against contamination a priority and foster coordinated management of interstate aquifers.

4. Increased attention must be given to prevention of groundwater contamination. Besides current efforts, we must develop alternatives to the contaminating activity, reduce waste hazard levels through the levy of taxes or fees, and restrict the manufacture, distribution, and use of contaminating substances. Only through aggressive preventive efforts can we minimize the expensive and often painful cleanup and remedial programs we are seeing in so many communities.

CHAPTER 2

Historical Development of the National Primary Drinking Water Regulations

Charles D. Larson

INTRODUCTION

History gives ample evidence of the inescapable penalties paid by past civilizations that failed to provide for the safety of their water systems. Modern history shows that such waterborne diseases as typhoid, dysentery, and cholera are controllable, and in fact were all but eliminated in the United States by the 1930s by applying the principles identified in what are commonly referred to as the U.S. Public Health Service (PHS) Drinking Water Standards. The National Community Water Supply Study[1] completed in 1969 suggested that we had begun to backslide, and the trend in waterborne disease shown in Figure 1[2,3] suggests the slide is not over, even though the Safe Drinking Water Act has been on the books since 1974. Waterborne disease outbreaks in Pennsylvania, Maine, and other locations, and organic chemical problems only partially foreseen by those involved in revising the 1962 PHS standards, have made safe, palatable drinking water an enormous challenge in the 1980s. The following history is based upon the writings of McDermott.[4]

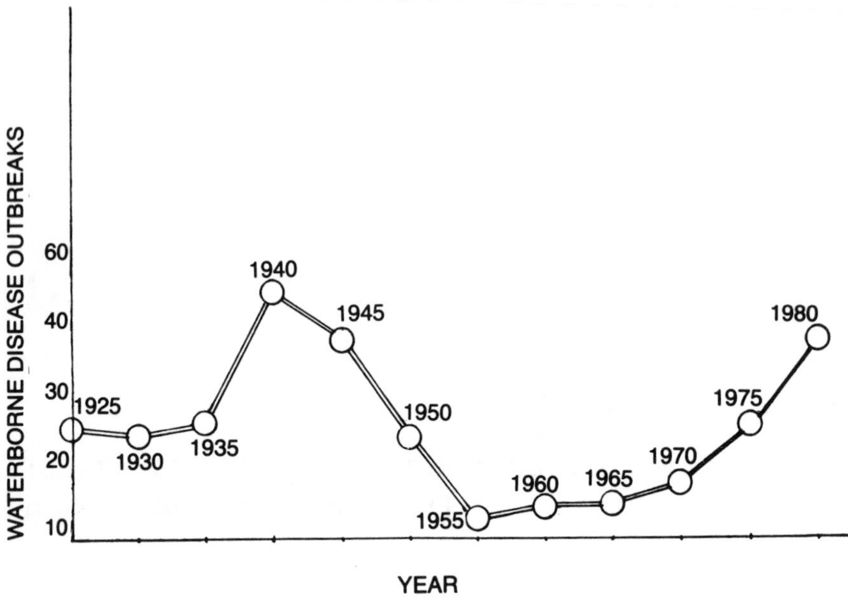

Figure 1. Average annual number waterborne disease outbreaks, 1925–1980.

HISTORICAL PERSPECTIVE

The first drinking water directives were issued in ancient times, although their scientific basis is of relatively recent origin. A Sanskrit source is quoted as saying at least 4000 years ago: "It is directed to heat foul water by boiling and exposing to sunlight and by dipping seven times into it a piece of hot copper, then to filter and cool in an earthen vessel. The direction is given by the God who is the incarnation of medical science."[5] Hippocrates, the father of medicine, identified the need for sanitary surveys 2400 years ago: "When one comes into a city to which he is a stranger,... one should consider most attentively the water which the inhabitants use, whether they be marshy and soft, or hard and running from elevated and rocky situations, whether it be naked and deficient."[4] Through the 1850s, it was a common belief in Western civilization that disease was caused by foul air. Though the epidemiologist Dr. John Snow was to prove in 1854 that cholera was a waterborne disease, the identification of specific disease-causing bacteria did not occur until the last quarter of the nineteenth century.[4]

Waterborne disease such as typhoid fever was prevalent in the United States then. McDermott, writing in 1973, states: "It can be noted that Pittsburgh reported 158 [typhoid] deaths per 100,000 persons during one year in the 1880s, while the Philadelphia experience varied from 60 to 80 per 100,000. Nationally, the typhoid death rate fell from 100/100,000 to about 30/100,000 by 1930 and is now insignificant as a direct result of the widespread use of the sand filtration process, the broad application of chlorine disinfection, and the issuance and application of drinking water standards."[4] Table 1 traces the development of drinking water standards with a brief summary of selected events.

Federal authority to establish standards for drinking water systems originated with the Interstate Quarantine Act of 1893. Among other things, this act authorized the Surgeon General of the U.S. Public Health Service "to make and enforce such regulations as in his judgment are necessary to prevent the introduction, transmission, or spread of communicable disease from foreign countries into the states or possessions, or from one state or possession into any other state or possession."[4] This provision resulted in promulgation of the Interstate Quarantine Regulations in 1894, and in 1912 the first water-related regulation: the use of the common drinking cup on interstate carriers was prohibited.[4]

1914 STANDARDS

"It was quickly realized," writes McDermott, "the most sanitary drinking cup would be of no value if the water placed in it was unsafe, and in 1914 the first official drinking water standard—a bacteriological standard—was adopted. It was made applicable to any system that provided water to an interstate common carrier. Such a water supply is called an Interstate Carrier Water Supply."[4] (At one time, there were over 2000 such systems. By 1986, the list had dwindled to 818: 635 public water supplies and 183 noncommunity water supplies.) The adoption of the drinking water standard stemmed from the need to have a common base for the determination of the sanitary condition of water systems and a foundation for action to protect the traveling public. From 1914 to 1975, federal, state, and local health authorities and waterworks officials used this standard to improve the nation's community water systems and to protect the public against waterborne disease.

1925 STANDARDS

McDermott writes: "Some people thought it was absurd to protect against a bacterially caused diarrhea while at the same time ignoring a chemically caused diarrhea such as can be induced by magnesium sulfate, a common contaminant of water. Hence, in 1925 the standards were

Table 1. An Abbreviated History of Drinking Water Standards

Year	Event
1853	F. Cohn in Germany used microscope and related water quality to algae and other microorganisms
1854	Dr. John Snow investigated cholera outbreak in London and determined it was related to contaminated drinking water
1885	Escherich discovered *Bacterium coli*
1891	Miquel established plate count standards for water quality
1893	Theobald Smith in New York used fermentation tubes for the first time
1893	Interstate quarantine regulations promulgated by the Secretary of Treasury
1903	Whipple introduced first standards for water quality using coliforms as indicators
1905	First standard methods published
1912	Common cup was banned on interstate carriers
1914	First U.S. Public Health Service (PHS) drinking water standards—*only bacteriological*
1925	First revision of PHS standards: Source protection Chemicals added Plate count dropped
1942	Second revision of PHS standards—the standards were separated into two parts: Standards with additional chemicals added Waterworks practice manual
1946	Third revision of PHS standards—waterworks practice manual published separately: Membrane filter was allowed in 1957
1962	Fourth revision of PHS standards: Waterworks practice part dropped CCE, ABS-detergents, barium, cadmium, cyanide, nitrate, and silver added Fluorides—climate considerations Radioactivity—included for the first time Included rationale used for chemical standards
1974	Safe Drinking Water Act (SDWA) signed into law on December 16, 1974
1975	National Interim Primary Drinking Water Regulations, promulgated by the Environmental Protection Agency, became effective June 24, 1977
1979	Trihalomethane regulations became effective: For population served: >75,000 – November 29, 1980 >10,000 – November 29, 1982
1986	Amendments to SDWA signed into law on June 19, 1986

Note: ABS = alkyl benzene sulfonate; CCE = carbon chloroform extract.

revised, and new sections on source and supply and physical and chemical characteristics (lead, copper, zinc, excessive soluble mineral substances) were added."[4]

1942 STANDARDS

In 1942, the standards were again revised. Writes McDermott: "One of the notable additions was a requirement that samples for bacteriological examination be obtained from points in the distribution system. Previously, bacteriological samples were taken only where the water left the water treatment plant, since this was the most convenient location."[4]

The other principal changes were:[4]

1. The text was divided into two parts, one containing the standards, the other a manual of waterworks practice. The latter was intended to serve as a guide to the reporting agency and was *not* to be considered additional requirements for certification of the water system.
2. A minimum number of samples were to be examined each month, and the laboratories and procedures used in making these examinations were subject to inspection at any time.
3. Maximum permissible concentrations were established for lead, fluoride, arsenic, and selenium. Salts of barium, hexavalent chromium, heavy metal glucosides, or other substances having deleterious physiological effects were not allowed in the water system.
4. Maximum concentrations that should not be exceeded where more suitable water supplies were available were set for copper, iron plus manganese, zinc, chloride, sulfate, phenolic compounds, total solids, and alkalinity.
5. A provision was added that the water supply system in all its parts should be free from sanitary defects and health hazards and should be maintained at all times in a proper sanitary condition.

1946 STANDARDS

The 1946 standards were essentially the same as the 1942 standards except for:[4]

1. The standards were made "generally acceptable and applicable to all water supplies in the United States."
2. "Manual of Recommended Water Sanitation Practice" was published separately. (This was done because, all too frequently, the manual had been interpreted as being part of the standards.)
3. A maximum permissible concentration was added for hexavalent chromium, and wording in the 1942 standards that prohibited the salts of

barium, hexavalent chromium, heavy metal glucosides, and other substances was changed to also prohibit their use in water treatment processes.

The 1946 standards were amended by publication in the *Federal Register* dated March 1, 1957, with provisions authorizing the use of the membrane filter procedure in the bacteriological examination of water samples.[4]

1962 STANDARDS

The objective of the advisory committee for the 1962 standards was to recommend minimum requirements for domestic water systems to protect the health and promote the well-being of individuals and the community.[4]

The major changes for the 1962 standards were:[4]

1. Water systems were required to be properly operated under the supervision of qualified personnel.
2. Permissible or recommended maximum limiting concentrations for alkyl benzene sulfonate (ABS), barium, cadmium, carbon chloroform extract (CCE), cyanide, nitrate, and silver were added.
3. In establishing the limiting levels of fluorides, the amount of water consumed by an individual as determined by climate was to be considered.
4. A new section on radioactivity was added.
5. An appendix explaining the rationale used by the committee in establishing the limiting concentration of various chemicals was added.

1963 ADVISORY COMMITTEE

In accordance with the recommendations made by the advisory committee for the 1962 standards, the Public Health Service, in 1963, established an advisory committee on the use of the Public Health Service Drinking Water Standards. This committee continued to appraise the quality requirements for drinking water and, in June 1967, recommended several changes and additions for inclusion in the Public Health Service Drinking Water Standards. The recommendations, based on toxicological concern, would have established maximum permissible concentrations for the following chemicals: aldrin, chlordane, DDT, dieldrin, endrin, heptachlor, heptachlor epoxide, lindane, methoxychlor, toxaphene, organic phosphates, carbamates, boron, and uranyl ion.

In attempting to promulgate these recommendations, however, someone recalled that the legislative base for the standards was restricted to the control of communicable disease. Following the reasoning that chemicals as such do not cause communicable disease, it was held that the Public Health Service had no authority to establish drinking water standards for chemicals. However, the foresight of the 1963 committee was not wasted, since in 1969, with the increased public concern about pesticide problems, these recommended limits were issued by the Division of Water Hygiene as guidelines for use by state and local health units as well as water resources planning and water pollution control agencies.

The increasing number of items that were either in or recommended for inclusion in the drinking water standards to protect the public from chronic illnesses associated with chemical contaminants indicated a mounting concern that human health might be endangered by toxic materials in water. This concern is not surprising when we consider the growth in the chemical industry that began in the 1930s (when waterborne disease of bacterial origin had been brought under control) and that a substantial part of the nation's chemical production can and does enter the waterways serving as drinking water sources. Some of the more provocative challenges to the nation's drinking water systems arise from such chemicals as nitrates, arsenic, selenium, mercury, organic carcinogens and teratogens, and trace metal antagonists.

PROPOSED 1973 STANDARDS THROUGH EPA INTERIM PRIMARY DRINKING WATER REGULATIONS

Because of difficulties encountered during the National Community Water Supply Study in applying the 1962 PHS standards, a federal technical committee was established in October 1969 by the PHS and charged with the responsibility of preparing necessary revisions and recommendations to revise the 1962 standards. The work of the committee was completed in December 1971, and among their recommendations was a public notification requirement. Their report was reviewed by an Environmental Protection Agency (EPA) advisory committee appointed by William D. Ruckelshaus, the first EPA Administrator. The report, made available in 1973, formed the basis for the Interim Primary Drinking Water Regulations promulgated in December 1975. Amended once before becoming effective in June 1977, these regulations were amended further in 1979, 1980, and 1982. These standards are summarized in Tables 2 through 7, with comparisons to previous editions of the PHS Drinking Water Standards. Table 8 presents a summary of the monitoring requirements for the current regulations.

The major change in these intervening years, beyond monitoring for sodium and corrosivity in 1980, was the addition of the total trihalo-

Table 2. Maximum Permissible Chemical Limits in U.S. Public Health Service (PHS) and U.S. Environmental Protection Agency (EPA) Drinking Water Standards

	PHS Standard				EPA NIPDWR
Chemical	**1925 (mg/L)**	**1942 (mg/L)**	**1946 (mg/L)**	**1962 (mg/L)**	**1975 (mg/L)**
Lead	0.1	0.1	0.1	0.05	0.05
Copper	0.2	a	a	a	—
Zinc	5.0	a	a	a	—
Fluoride	—	1.0	1.5	b	4.0[c]
Arsenic	—	0.05	0.05	0.05	0.05
Selenium	—	0.05	0.05	0.01	0.01
Hexavalent chromium	—	0.0	0.05	0.05	0.05
Barium	—	—	—	1.0	1.0
Cadmium	—	—	—	0.01	0.01
Cyanide	—	—	—	0.2	—
Silver	—	—	—	0.05	0.05
Nitrate	—	—	—	—	10 as N
Mercury	—	—	—	—	0.002

Source: Based on PHS provisions.[6]

Note: Dash = not included; NIPDWR = National Interim Primary Drinking Water Regulations.

[a] Change to recommended limits after 1925.

[b] Limits for naturally occurring and supplementation of fluoride based on table of annual average of maximum daily air temperatures.

[c] *Federal Register*, April 2, 1986; effective May 2, 1986.

methane regulation in 1979. Table 9 presents the regulation in outline form. One question that has been raised many times about the trihalomethane regulation is why systems serving fewer than 10,000 people were not covered by the regulation. (It should be noted that the states do have the option of applying the regulation to smaller systems.) The EPA considered the following in making its decision:

1. that about 80% of these systems use groundwater and therefore usually have low levels of precursor substances like humic acid
2. that not as many of these systems use chlorine
3. that these systems usually have smaller distribution systems, which means shorter transport times and thus less contact with chlorine, if it is used
4. EPA's ability to administer the extra 57,000 systems
5. the availability of laboratory resources
6. that smaller systems tend to have greater microbiological risk, and changes in treatment to reduce THMs might compromise disinfection

Table 3. Recommended Chemical Limits in U.S. Public Health Service (PHS) and U.S. Environmental Protection Agency (EPA) Drinking Water Standards.

	PHS Standard				EPA NISDWR
Chemical	**1925 (mg/L)**	**1942 (mg/L)**	**1946 (mg/L)**	**1962 (mg/L)**	**1979 (mg/L)**
Copper	—	3.0	2.0	1.0	1.0
Zinc	—	15.0	15.0	5.0	5.0
Iron	0.3	0.3	0.3	0.3	0.3
Manganese	—	—	—	0.1	0.05
Magnesium	100	125	125	—	—
Chloride	250	250	250	250	250
Sulfate	250	250	250	250	250
Phenols	—	0.001	0.00	0.001	—
Total Solids	1000	500[a]	500[a]	500	500
ABS	—	—	—	0.05	0.051
Arsenic	—	—	—	0.01	—
CCE	—	—	—	0.2	—
Cyanide	—	—	—	0.01	—
Nitrate	—	—	—	45.0	—
Fluoride	—	—	—	—	2.0[b]
pH	—	—	—	—	6.5–8.5
Color	—	—	—	—	15 color units
Corrosivity	—	—	—	—	Noncorrosive
Odor	—	—	—	—	3 TON

Source: Based on PHS provisions.[6]

Note: Dash = not included; ABS = alkyl benzene sulfonate; CCE = carbon chloroform extract; NISDWR = National Interim Secondary Drinking Water Regulations; TON = threshold odor number.

[a] Total solids of 1000 mg/L may be permitted.

[b] *Federal Register*, April 2, 1986; effective May 2, 1986.

Table 4. Bacteriological Requirements of U.S. Public Health Service (PHS) Drinking Water Standards Since 1914

Standard	**Coliform Organisms Per 100 mL**	**Total Plate Count, 37°C (org./mL)**
1914	2.2	100
1925	<1.05	No requirements
1942	<2.2	No requirements
1946	<2.2	No requirements
1962	<2.2	No requirements

Source: Based on PHS provisions.[6]

7. that about 80% of the population served by public systems are covered by present regulations

Another question regarding the trihalomethane regulation was: What is the "best available treatment" that must be used to meet the regulation? In the February 28, 1983 *Federal Register,* starting on page 8413, EPA defines *best technology, techniques, or other means generally available.* The following items would be required for compliance with the regulation:

Table 5. Maximum Permissible Microbiological Contaminants (NIPDWR)

Coliform Method	Per Month	Less than 20 Samples per Month	20 or More Samples per Month
	Number of coliform bacteria shall not exceed:		
Membrane filter (100-mL portions)	1/100 mL average density	4/100 mL in one sample	4/100 mL in 5% of samples
	Coliform bacteria shall not be present in more than:		
Multiple tube fermentation (10-mL portions)	10% of portions	3 portions in one sample	3 portions in 5% of samples
Coliform Method	**Per Month**	**Less than 5 Samples per Month**	**5 or More Samples per Month**
	Coliform bacteria shall not be present in more than:		
Multiple tube fermentation (100-mL portions)	60% of portions	5 portions in more than one sample	5 portions in more than 20% of samples

Source: Based on EPA training course data.[7]

Note: NIPDWR = National Interim Primary Drinking Water Regulations.

Table 6. Maximum Permissible Radioactivity Limits

	PHS Standard				EPA NIPDWR
Radionuclides	1925	1942	1946	1962 (pCi/L)	1975 (pCi/L)
Radium[a]	—	—	—	3	5
Strontium-90	—	—	—	10	—
Gross alpha activity	—	—	—	1000	—
Gross alpha activity[b]	—	—	—	—	15

Source: Based on PHS provisions.[6]

Note: Dash = not included; EPA = U.S. Environmental Protection Agency; NIPDWR = National Interim Primary Drinking Water Regulations; PHS = U.S. Public Health Service.

[a] Combined radium-226 and radium-228.

[b] Including radium-226 but excluding radon and uranium.

Table 7. Maximum Permissible Levels of Pesticides in Drinking Water

Pesticide	PHS Guidelines—1969 (mg/L)	EPA Proposed Standards—1973 (mg/L)	EPA NIPDWR—1975 (mg/L)
Chlorinated hydrocarbons			
Aldrin	0.017	0.001	—
Chlordane	0.003	0.003[a]	—
DDT	0.042	0.05	—
Dieldrin	0.017	0.001	—
Endrin	0.001	0.0005	0.0002
Heptachlor	0.018	0.0001	—
Heptachlor epoxide	0.018	0.0001	—
Lindane	0.056	0.005	0.004
Methoxychlor	0.035	1.0	0.1
Toxaphene	0.005	0.005[a]	0.005
Chlorophenoxys			
2,4-D	—	—	0.1
2,4,5-TP Silvex	—	—	0.01

Note: Dash = not included; EPA = Environmental Protection Agency; NIPDWR = National Interim Primary Drinking Water Regulations; PHS = U.S. Public Health Service.

[a] Limit selected on basis of odor, although the toxic limits were similar.

Table 8. Summary of IPDWR Monitoring Requirements

	Surface Source	Ground Water
Coliform[a]	Number based on population	Same as surface water, but state may reduce to 1/quarter
Inorganic[b]	1 sample per year at free-flowing tap	Sampling at 3-year intervals
Pesticides	Sample at 3-year intervals	Only if state requires sampling
Natural radioactivity	4-year intervals	Within 3 years of effective date; then at 4-year intervals
Sodium	Sampling and analysis annually	Sampling and analysis every 3 years
Corrosivity characteristics	Two samples and analysis: one midsummer one midwinter	One sample and analysis
Turbidity	Once per day at entrance to distribution system	Not applicable
TTHMs	4 samples/quarter	4 samples/quarter
>10,000 population	May be reduced to 1/quarter	May be reduced to 1/quarter

Source: Based on EPA training course data.[7]

Note: IPDWR = Interim Primary Drinking Water Regulations; TTHMs = total trihalomethanes.

[a] Keep records for 5 years.

[b] Keep records for 10 years.

Table 9. Summary of Total Trihalomethane (TTHM) Regulations

Maximum contaminant level (MCL) = 0.10 mg/L (100 μg/L) total trihalomethanes

Applies to community water systems that add a disinfectant to the treatment process (ground- or surface water)

Effective dates:
- Systems >75,000 – Nov. 29, 1981
- Systems 10-75,000 – Nov. 29, 1983
- Systems <10,000 – state discretion

Sampling:
- 4 samples per quarter = 16 per year
- 25% at extreme of distribution system
- 75% to represent population distribution

Frequency can be reduced to 1 per quarter by state upon review of data or maximum total trihalomethane potential (MTP)—one MTP per year if it is less than 0.1 ppm

Report: Average of quarterly analysis to state within 30 days

Compliance: Running average of 1 yr of data must be less than MCL

Source: Based on EPA training course data.[7]

1. improved clarification for THM precursor reduction
2. use of chloramines as an alternate or supplemental disinfectant or oxidant
3. use of chlorine dioxide as an alternate or supplemental disinfectant or oxidant
4. moving the point of chlorination to reduce TTHM formation
5. use of powdered activated carbon for THM precursor reduction or TTHM reduction

In addition, the methods listed below might be required:

1. introduction of off-line storage for THM precursor reduction
2. use of aeration for TTHM reduction where geographically and environmentally appropriate
3. installation of clarification where not currently practiced
4. consideration of alternative sources of raw water
5. use of ozone as an alternate or supplemental disinfectant or oxidant

The use of granular activated carbon or biologically activated carbon will not be required to meet the TTHM regulation. To help states, the EPA has issued guidance to be followed when treatment is modified, to prevent compromises in disinfection. In the chapters that follow, you will be brought up to date on the compliance with these regulations and the SDWA Amendments signed into law on June 19, 1986.

REFERENCES

1. "Community Water Supply Study; Significance of the Study," Bureau of Water Hygiene, U.S. Public Health Service, July 1970.
2. Craun, G. F., and L. J. McCabe. "Review of the Causes of Waterborne Disease Outbreaks," *J. AWWA* 65(1):74–84 (1973).
3. Lippy, E. C. and S. C. Waltrip. "Waterborne Disease Outbreaks—1946–1980: A Thirty-Five Year Perspective," *J. AWWA* 76(2):60–67 (1984).
4. McDermott, H. "Federal Drinking Water Standards—Past, Present and Future," *Water Well J.* 27(12):29–35 (1973). Also, *J. Environ. Eng. Div. ASCE* 99(EE4):469–478 (1973).
5. Baker, M. N. *The Quest for Pure Water* (New York: The American Water Works Association, 1949).
6. "Provisions of the 1962 Public Health Service Drinking Water Standards," from Interstate Carrier Water Supply Program, Environmental Engineering and Food Protection Services, Public Health Service, Region VII, Dallas, TX. Given to participants of Texas Water & Sewage Works Association Short School, March 12–16, 1962.
7. "Sanitary Survey Training Course Manual," Office of Drinking Water, U.S. Environmental Protection Agency, Washington, DC, October, 1983.

SUGGESTED READING

Taylor, F. B. "Drinking Water Standards Principals and History 1914–1976," *J. NEWWA* 91(4):237–259 (1977).

CHAPTER 3

Overview of the Current National Primary Drinking Water Regulations and Regulation Development Process

Joseph A. Cotruvo and Marlene Regelski

INTRODUCTION

The Safe Drinking Water Act of 1974 (SDWA) directed the U.S. Environmental Protection Agency (EPA) to identify substances in drinking water "which in the judgment of the administrator may have any adverse effect" on public health.[1] Interim National Primary Drinking Water Regulations (INPDWR) were established as directed within 180 days of enactment of the SDWA and final regulations were to be developed over a specified period of years. The schedule is shown in Table 1.

The promulgation of the National Primary Drinking Water Regulations (NPDWR) follows specific steps. First, the Advance Notice of Proposed Rule Making (ANPRM) is published. It includes what EPA is considering for possible regulation and what is already known about the science and technology of the contaminant, and asks for more information and comments (see Figure 1). Second, the EPA, as mandated by the SDWA Amendments,[2] proposes maximum containment levels (MCLs), (enforceable standards) and maximum contaminant level goals (MCLGs) (nonenforceable health goals) simultaneously. More public comments are received, and eventually the regulations are finalized.

Table 1. Safe Drinking Water Act Compliance Deadlines

Substances	Statutory Deadlines
9 MCLGs and MCLs + monitoring	June 19, 1987
Public notice revisions	Sept. 19, 1987
Filtration criteria	Dec. 19, 1987
Monitoring for unregulated contaminants	Dec. 19, 1987
List of contaminants (DWPL)—final	Jan. 1, 1988
40 MCLGs and MCLs + monitoring	June 19, 1988
34 MCLGs and MCLs + monitoring	June 19, 1989
Disinfection treatment	June 19, 1989
25 MCLGs and MCLs + monitoring	Jan. 1, 1991

Note: DWPL = drinking water priority list; MCL = maximum contaminant level; MCLG = maximum contaminant level goal.

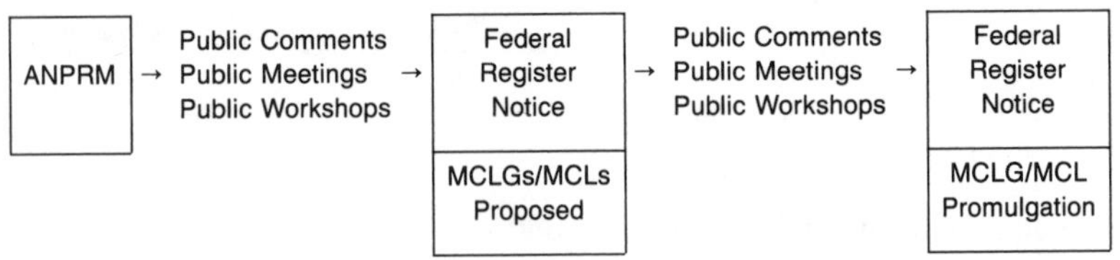

Figure 1. Regulatory development process.

Table 2. Office of Drinking Water Regulatory Phase Schedule

Phase	Substance	Expected Promulgation Date
I	Volatile organic chemicals (VOCs)	July 8, 1987
II	Synthetic organic chemicals (SOCs) and inorganic chemicals (IOCs)	June 1989
	Microbials and surface water treatment (filtration)	June 1989
	Lead/Copper (corrosion by-products)[a]	December 1988
III	Radionuclides	December 1988
IV	Disinfectants and disinfection by-products	June 1989
V	Other inorganic chemicals, synthetic organic chemicals, and pesticides	June 1989
VI	25 additional chemicals	January 1991

[a] Have been separated from Phase II. Proposed NPDWR published.

The Office of Drinking Water developed a six-phase schedule that has attempted to parallel the SDWA-specified deadlines (see Table 2). Each phase of the six-phase regulatory program as well as other priorities set by the SDWA such as compliance requirements, surface water treatment and disinfection criteria, variances and exemptions, and regulatory timetables and deadlines will be discussed in detail in Chapters 4 through 9 of this book. In this chapter we will summarize the directives of the Safe Drinking Water Act Amendments of 1986, describe the philosophy of and some of the methods used by the EPA in developing the drinking water regulations, and define some of the terms used in the regulations.

DIRECTIVES OF THE SDWA AMENDMENTS

- EPA is to set MCLGs and NPDWRs (in the form of MCLs and treatment processes) for 83 specific contaminants and for any other contaminant in the drinking water which may have any adverse effect upon the health of persons and which is known or anticipated to occur in public water systems (see Table 3).
- EPA is to set regulations for a specific list of contaminants (SDWA of 1974):

9 MCLs in 12 months	(June 19, 1987)
40 MCLs in 24 months	(June 19, 1988)
34 MCLs in 36 months	(June 19, 1989)

  A list of seven substitutes is allowed if regulation of any seven other contaminants would be more protective of public health, and other contaminants would be listed for public comment.
- RMCL (recommended maximum contaminant level) terminology is changed to MCLG (maximum contaminant level goal).
- MCLs are to be set as close to MCLGs as is feasible. The term *generally available* (technology) was changed to *as is feasible. Feasible* is defined as "with the use of the best technology, treatment techniques and other means, which the Administrator finds, after examination for efficacy under field conditions and not solely under laboratory conditions, are available (taking cost into consideration)."
- Granular activated carbon (GAC) is stated to be feasible for the control of synthetic organic chemicals (SOCs), and any technology or other means found to be the best available for control of SOCs must be at least as effective as GAC.

Table 3. Contaminants Required to be Regulated Under the Safe Drinking Water Act Amendments of 1986

Volatile Organic Chemicals	Nitrate	Epichlorohydrin
Trichloroethylene[a]	Selenium	Toluene
Tetrachloroethylene	Silver	Adipates
Carbon tetrachloride[a]	Fluoride[c]	2,3,7,8-TCDD (Dioxin)
1,1,1-Trichloroethane[a]	Aluminum	1,1,2-Trichloroethane
1,2-Dichloroethane[a]	Antimony	Vydate
Vinyl chloride[a]	Molybdenum	Simazine
Methylene chloride	Asbestos	Polyaromatic hydrocarbons (PAHs)
Benzene[a]	Sulfate	Polychlorinated biphenyls (PCBs)
Monochlorobenzene	Copper	Atrazine
Dichlorobenzene[b]	Vanadium	Phthalates
Trichlorobenzene	Sodium	Acrylamide
1,1-Dichloroethylene[a]	Nickel	Dibromochloropropane (DBCP)
trans-1,2-Dichloroethylene	Zinc	1,2-Dichloropropane
cis-1,2-Dichloroethylene	Thallium	Pentachlorophenol
	Beryllium	Pichloram
Microbiology and Turbidity	Cyanide	Dinoseb
Total coliforms		Ethylene dibromide (EDB)
Turbidity	**Organics**	Dibromomethane
Giardia lamblia	Endrin	Xylene
Viruses	Lindane	Hexachlorocyclopentadiene
Standard plate count	Methoxychlor	
Legionella	Toxaphene	**Radionuclides**
	2,4-D	Radium-226 and -228
Inorganics	2,4,5-TP	Beta particle and photon radioactivity
Arsenic	Aldicarb	Radon
Barium	Chlordane	Gross alpha particle activity
Cadmium	Dalapon	Uranium
Chromium	Diquat	
Lead	Endothall	
Mercury	Glyphosphate	
	Carbofuran	
	Alachlor	

Note: MCL = maximum contaminant level.

[a] Promulgated July 8, 1987

[b] MCL for *p*-dichlorobenzene was published July 8, 1987; *ortho*-dichlorobenzene is on additional list for consideration

[c] Promulgated April 2, 1986

- MCLGs and MCLs are to be proposed and promulgated simultaneously.
- MCLGs and MCLs/monitoring requirements are to be set for 83 contaminants listed in the SDWA. The best available technology (BAT) is also to be specified for each.
- EPA must establish a drinking water priority list of contaminants (DWPL) (see Table 4). (The proposed list was published July 8, 1987;[3] final notice was published in December 1987.)

Table 4. Drinking Water Priority List

Zinc	2,4-Dinitrotoluene
Silver	1,3-Dichloropropane
Sodium	Bromobenzene
Aluminum	Chloromethane
Molybdenum	Bromomethane
Vanadium	1,2,3-Trichloropropane
Dibromomethane	1,1,1,2-Tetrachloroethane
Chlorine	Chloroethane
Hypochlorite ion	2,2-Dichloropropane
Chlorine dioxide	*o*-Chlorotoluene
Chlorite	*p*-Chlorotoluene
Chloramine	Hexachlorobenzene
Ammonia	Hexachloroethane
Trihalomethanes (chloroform, dibromochloromethane, bromodichloromethane, bromoform)	Hexachlorobutadiene
	1,1-Dichloropropene
	2,4,5-T
Chlorophenols	Isophorone
Halonitriles	Ethylene thiourea
Selected disinfection-related chlorinated acids, alcohols, aldehydes, and ketones	Boron
	Strontium
	Cryptosporidium
Chloropicrin	

- MCLs and MCLGs must be set for at least 25 contaminants on the list by January 1, 1991, then 25 more every three years hence.
- Criteria for placement on DWPL is described:
  -occurrence or potential occurrence in drinking water
  -adverse health effects
  -sufficient data available to set an MCLG and MCL

  The chemicals on the DWPL were taken from a number of sources: the seven contaminants taken off the original list of 83, disinfectants and disinfection by-products, the first 50 contaminants specified under Section 110 of the Superfund Amendments and Reauthorization Act of 1986 (SARA), pesticides included as design-analytes in the National Pesticide Survey (NPS), volatile organic chemicals (VOCs) reported in Section 1445 of SDWA as unregulated VOCs to be monitored, and certain other substances reported frequently and/or occurring at high concentrations in other recent surveys.

- Criteria for selection onto the DWPL contaminant priority list is described:
  -The contaminant must occur in public water systems, or its characteristics or use patterns must be such that it has a strong potential to occur in public water systems at levels of concern.

-The contaminant must have a documented or suspected adverse human health effect.

-There must be sufficient information available on the contaminant so that a regulation could be developed within the statutory time frames.

Substances for which information is insufficient to develop regulations will be candidates for subsequent priority lists (to be published triennially beginning in 1991.) (Further information on the specific selection criteria may be found in the *Federal Register,* 52 FR 25720[4] and 53 FR 1892.[5])

- Candidates for substitution onto "list of 83" are given:

  -Aldicarb sulfone
  -Aldicarb sulfoxide
  -Ethylbenzene
  -Heptachlor
  -Heptachlor epoxide
  -Styrene
  -Nitrite

  The proposed list of seven substitutes and replacements on the original list of 83 was published in the *Federal Register* on July 8, 1987 (52 FR 25720). The final list was published January 22, 1988 (53 FR 1892).

- Candidates for removal from the "list of 83" are given:

  -Aluminum
  -Dibromomethane
  -Molybdenum
  -Silver
  -Sodium
  -Vanadium
  -Zinc

  The seven chemical candidates listed for substitution onto the "list of 83" will now be placed on the DWPL (Table 4).

- Three other candidates considered for removal from the "list of 83" are listed for public comment:

  -Phthalates
  -Sulfate
  -1,1,2-Trichloroethane

- Monitoring requirements are to be set to assure compliance with the MCLs. In most cases, states have the responsibility for enforcement of MCLs. Public notification of a violation by a public water system of an MCL or monitoring requirement is required to be given.

PROCESS OF STANDARDS DEVELOPMENT

The standards development process involves an intensive technological evaluation that includes assessments of:

- occurrence in the environment
- human exposure in specific and general populations
- adverse health effects
- risks to the population
- methods of detection
- chemical transformations of the contaminant in drinking water
- treatment technologies and costs

In selecting contaminants for regulation, the most relevant criteria are (1) potential health risk, (2) ability to detect a contaminant in the drinking water, and (3) occurrence or potential occurrence in drinking water.[6]

For each of the substances or contaminants the EPA selects, there are two methods for development of regulatory measures: either the EPA must establish a maximum contaminant level (MCL), or if it is not economically or technically feasible to monitor the contaminant level in the drinking water, a treatment method to remove the contaminant from the water supply must be specified.

MCLs and MCLGs

The SDWA directs EPA to set MCLs as close to MCLGs as feasible[7] (see Figure 2). *Feasible* means as close as possible "with the use of best technology, treatment techniques, or other means which the Administrator finds available (taking cost into consideration) after examination for efficacy under field conditions and not solely under laboratory conditions." MCLGs, previously called recommended maximum contaminant levels (RMCLs), are set at concentration levels at which no known or anticipated adverse human health effects would occur and which allow an adequate margin of safety (MOS).

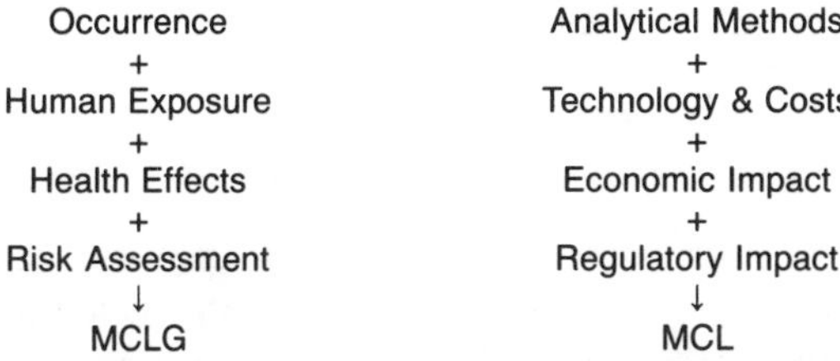

Figure 2. MCL/MCLG development.

Noncarcinogens

For noncarcinogens, MCLGs are derived from the reference dose (RfD)—formerly called acceptable daily intake (ADI)—calculated for each specific contaminant.

The RfD is an estimate, with an uncertainty factor expressed in orders of magnitude, of a daily exposure to the human population (including sensitive subgroups) that is likely to be without an appreciable risk of deleterious health effects during a lifetime. To derive the RfD, a no or lowest observed adverse effect level (NOAEL OR LOAEL) that has been identified from a subchronic or chronic scientific study of humans or animals is divided by the uncertainty factor. The uncertainty factor is calculated to take care of intra- and interspecies variations, limited or incomplete data, significance of the adverse effect, and pharmacokinetic factors.

From the RfD, a drinking water equivalent level (DWEL) is calculated by multiplying the RfD by an assumed body weight of 70 kg for an adult and then divided by an average daily water consumption (for an adult) of 2 liters per day. The DWEL assumes there will be a 100% exposure to a substance from drinking water and provides the noncarcinogenic health effect bases for the MCLG. Further, the MCLG is determined by multiplying the DWEL by the percentage of the total daily exposure contributed by drinking water (relative source contribution), as seen in the equation below:

$$\text{MCLG} = \frac{(\text{NOAEL})\ (70\ \text{kg})\ (\%\ \text{contribution drinking water})}{(\text{Uncertainty Factor})\ (2\ \text{L}\ H_2O/\text{day})}$$

where NOAEL is expressed in mg/kg/day
70 kg = assumed weight of an adult

For noncarcinogens, the MCLG will often equal the MCL.

Carcinogens

Where the contaminant is considered a known or probable human carcinogen (potentially without a threshold no-effect level) the MCLG is set at zero.

Potential Carcinogens

A separate health assessment system is used for potentially nonthreshold, no-effect-level chemicals with carcinogenic potential.

Non-threshold effect means that it is scientifically impossible to demonstrate experimentally that a dose can be found under which there is no effect. In other words, any amount of exposure is assumed to represent some finite level of risk for carcinogenicity in the absence of sufficient negative information. In the House Report[8] which accompanied the SDWA of 1974, precedence was given to this method, which suggested that recommended maximum contaminant levels (RMCLs, now MCLGs) for nonthreshold toxicants (i.e., carcinogens) should be zero.

Assessment for nonthreshold toxicants is made up of the weight of evidence scheme of carcinogenicity in humans (i.e., bioassays in animals and human epidemiological studies), as well as information that provides indirect evidence (i.e., mutagenicity and other short-term test results). The objectives of the assessment are 1) to determine the level or strength of evidence that the substance is a human or animal carcinogen, and 2) to provide an upper-bound estimate of the possible risks of human exposure to the substances in drinking water.

Classification of Carcinogenic Chemicals

Several scientific groups, including EPA, National Academy of Sciences Safe Drinking Water Committee (NAS), and the International Agency for Research on Cancer (IARC),[9] have designed carcinogenic chemical classification schemes on the basis of weight of available evidence for carcinogenicity. A summary of these schemes follows:

National Academy of Sciences (NAS)—four categories:

1 Human carcinogen
2 Suspected human carcinogen
3 Animal carcinogen
4 Suspected animal carcinogen

International Agency for Research on Cancer (IARC)—three groups:

1 Chemical is carcinogenic to humans based on sufficient epidemiological evidence.
2 Chemical is probably carcinogenic to humans based on limited human data (2A) or sufficient animal data and inadequate human data (2B).
3 Chemical cannot be classified as to its carcinogenicity in humans.

Environmental Protection Agency (EPA)—five groups:

A Human carcinogen based on sufficient evidence from epidemiological studies
B Probable human carcinogen based on at least limited evidence of carcinogenicity to humans (B_1), or usually a combination of sufficient evidence in animals and inadequate data in humans (B_2)
C Possible human carcinogen based on limited evidence of carcinogenicity in animals in the absence of human data

D Not classified based on inadequate evidence of carcinogenicity from animal data

E No evidence of carcinogenicity for humans (no evidence for carcinogenicity in at least two adequate animal tests in different species or in both epidemiological and animal studies

The EPA has decided to use the classification schemes as shown in Table 5. The MCLG for Category I chemicals is set at zero. A zero level is below the analytical detection level and is often unattainable. MCLGs for Category II contaminants take into consideration the reference dose (RfD) approach, tagging on the margin of safety (MOS) or orders of magnitude for the cancer risk range of the population (i.e., 10^{-5} or 10^{-6}). Category III adheres to the NOAEL approach to accommodate the extrapolation of animal data to human risk factors, the existence of weak or insufficient data, and individual differences in human sensitivity to toxic agents.

General guidelines based on the NAS recommendations regarding the use of calculated uncertainty factors to improve the margins of safety are as follows:[10]

10—Used with valid experimental results on appropriate durations of exposure in humans

100—Used when human data are not available and when extrapolating from valid results of long-term animal studies

1000—Used when human data are not available and when extrapolating from animal studies of less than chronic exposure

1-10—Additional safety factor using a LOAEL instead of a NOAEL

Table 5. EPA's Three-Category Approach

Category	Evidence of Carcinogenicity	Class	MCLG Setting Approach
I	Strong evidence	EPA group A or B	Zero
II	Equivocal evidence	EPA group C	1. RfD approach with additional safety factor, or 2. 10^{-5} to 10^{-6} cancer risk range
III	Inadequate or no evidence from animal data	EPA group D or E	RfD approach

Note: EPA = Environmental Protection Agency; MCLG = maximum contaminant level goal; RfD = reference dose.

Intermediate uncertainty factor—Other uncertainty factor used according to scientific judgment when justified

Treatment Methods

Treatment technologies are included as part of the regulatory development for each chemical or group of chemicals. Available treatment technologies and analytical methods are part of the analysis of the regulatory and economic consequences considered for each contaminant. Parameters of how to monitor, measure, and treat for a specific contaminant or a mixture of contaminants must be included in—in fact, be an integral part of—any standard that is promulgated. Regulations must say specifically what is the best available technology (BAT) for treatment procedures.

Analytical Methods

For purposes of treatment and monitoring of compliance, EPA must specify the analytical method best suited to detect the amount of a contaminant in drinking water. Because setting an MCL or MCLG below the smallest chemical amount detectible is not feasible for each MCL or MCLG derived, an accompanying analytical method, such as purge and trap gas or high-performance liquid chromatography, mass spectrometry, photoionization, or others, is specified.

Feasibility Studies and Laboratory Standards

To determine feasibility of controlling contaminants (i.e., accurately determining compliance) requires an evaluation of (1) availability and cost of analytical methods, (2) availability and performance of technologies (and other factors relative to feasibility), identifying those that are best, and (3) costs of the application of technologies to achieve various concentrations.

One of the measurements used in setting the laboratory performance requirements for an MCL is called the method detection limit (MDL), which is based upon the strictest laboratory specifications. The MDL is the minimum concentration of a substance that can be measured and reported with 99% confidence that the true value is greater than zero. These MDLs are measured by a few of the most experienced labs under nonroutine and controlled research type conditions.

A second measurement used by EPA, the practical quantitation level (PQL), is not lab- or time-specific but can provide a uniform concentration measurement that can be used to set standards. The PQL is the lowest measurement level that can be reliably achieved within specified limits of precision and accuracy during routine laboratory operating conditions. They are based on four factors:

1. quantitation
2. precision and accuracy
3. expected normal laboratory operations
4. the fundamental need (in the compliance and monitoring program) to have a sufficient number of laboratories available to conduct analyses

PQLs set a target performance for laboratories and provide consistency in implementing a regulatory program in a practical way, where both quality control and quality assurance are critical in a large number of laboratories.

Other key factors in the analysis of feasibility include:

- levels of chemical concentrations in drinking water supplies
- feasibility and/or reliability of removing contaminants to specific concentrations
- concentrations attainable by application of best technology generally available
- costs of treatment to achieve contaminant removal
- other factors, such as air pollution and waste disposal and their effects on other drinking water quality parameters

REFERENCES

1. *The Safe Drinking Water Act,* 42 U.S.C., 300f et seq. (1974).
2. *The Safe Drinking Water Act Amendments of 1986,* P.L. 99–339 (1986).
3. *Federal Register,* Vol. 52, No. 130, p. 25733 (1987).
4. *Federal Register,* Vol. 52, No. 130, p. 25720 (1987).
5. *Federal Register,* Vol. 53, No. 14, p. 1892 (1988).
6. *Federal Register,* Vol. 50, No. 219, p. 46941. (1985).
7. Cotruvo, J. A., S. Goldhaber, and C. Vogt. "Development of Drinking Water Regulations for Organic Contaminants in the U.S.," U.S. EPA paper presented at 2d National Conference on Drinking Water, Edmonton, Alberta, p. 8. (1986).
8. "House of Representatives Report No. 1185," 93rd Congress, 2d Session, 20 (1974).
9. *Federal Register,* Vol. 49, p. 46294.
10. *Federal Register,* Vol. 50, No. 219, p. 46946 (1985).

CHAPTER 4

National Primary Drinking Water Regulations for Volatile Organic Chemicals

Joseph A. Cotruvo and Marlene Regelski

INTRODUCTION

On July 8, 1987, the final rule was published for the National Primary Drinking Water Regulations (NPDWRs) in volatile organic chemicals (VOCs) and monitoring for unregulated contaminants.[1] The setting of maximum contaminant levels (MCLs) for the eight VOCs in this rule* (plus fluoride, promulgated April 6, 1986), satisfied the statutory deadlines in the Safe Drinking Water Act of 1974 (SDWA) in regard to completing the establishment of the first nine MCLs within 12 months.

The eight synthetic VOCs in this rule are widely used as unleaded gas additives; household cleaning solutions; solvents for removing grease from clothes, electronics and aircraft engines; air fresheners; and mothballs. They are found frequently in drinking water from groundwater sources. All have relatively low boiling points and vaporize readily.

*Vinyl chloride, benzene, trichloroethylene, carbon tetrachloride, 1,2-dichloroethane, *para*-dichlorobenzene, 1,1-dichloroethylene, and 1,1,1-trichloroethane.

BASIS FOR THE MCLs

The MCLs proposed by the U.S. Environmental Protection Agency (EPA) for the eight VOCs were based upon an evaluation of 1) availability and performance of treatment technologies, 2) availability, performance, and cost of analytical methods, and 3) costs of application of various technologies to lower the concentration of VOCs in drinking water to various levels. In reviewing the different technologies available for VOC removal, EPA considered the following factors: removal efficiency, degree of compatibility with other treatment processes, service life, and the ability to achieve compliance for all the water in a public water system. Based on these criteria, EPA proposed granular activated carbon (GAC) and packed tower aeration (PTA) as best available technologies (BAT) for all VOC removal (except vinyl chloride, for which only PTA is designated BAT). These technologies have 90–99% removal efficiency, are commercially available, and have been used successfully to remove VOCs in groundwater from both influents and effluents in many locations throughout the United States.

FINAL MCLs

The final MCLs for the eight VOCs (taken from the first part of Table 4 in Chapter 3) are categorized in Table 1. Note that for all the chemicals with zero maximum contaminant level goals (MCLGs) (except vinyl chloride), the MCLs are set at 0.005 mg/L. This number represents the "feasible" level, taking cost into consideration. With an MCL of 0.005 mg/L, about 1300 community water systems (CWSs), with a total capital cost of $280 million, need to install treatment capabilities to satisfy the requirements.

The MCL for vinyl chloride, however, at 0.002 mg/L does not incur any increased cost over the 0.005 mg/L level. Very few, if any, public water systems would have to install treatment solely to control vinyl chloride. Systems with that chemical's presence at any level virtually always have one or more of the other VOCs present, since vinyl chloride is known to appear as a degradation product of PCE or TCE.

The same treatment method, packed tower aeration, will remove vinyl chloride to a 0.002 mg/L level. Although this level may be harder to accurately measure, EPA recognizes that vinyl chloride is a known human carcinogen of possibly higher potency, and the risk posed by each unit of exposure could be higher than the equivalent unit of any other one of the four zero-MCLG VOCs.

para-Dichlorobenzene was originally classified as a Class D with a proposed MCL/MCLG of 0.75 mg/L. However, during the last two

Table 1. VOCs: Final MCLGs and MCLs

Contaminant	Health Effect	EPA Class	Final MCLG[a] (mg/L)	Final MCL
Vinyl chloride	Known human carcinogen	(A)	zero	0.002
Benzene	Known human carcinogen	(A)	zero	0.005
Trichloroethylene	Probable carcinogen	(B)	zero	0.005
Carbon tetrachloride	Probable carcinogen	(B)	zero	0.005
1,2-Dichloroethane	Probable carcinogen	(B)	zero	0.005
para-Dichlorobenzene	Possible carcinogen	(C)	0.075	0.075
1,1-Dichloroethylene	Possible carcinogen	(C)	0.007	0.007
1,1,1-Trichloroethane	Causes liver, circulatory system and central nervous system (CNS) damage	(D)	0.2	0.2

Note: EPA = Environmental Protection Agency; MCL = maximum contaminant level; MCLG = maximum contaminant level goal; VOC = volatile organic chemical.

[a] Final MCLGs were published Nov. 13, 1985. The MCLG and MCL for *p*-dichlorobenzene were reproposed at zero and 0.005 mg/L on April 17, 1987; comment was requested on levels of 0.075 mg/L and 0.075 mg/L, respectively.

years new studies have become available that show differing evidence. On April 19, 1987, a reproposal was submitted suggesting an upgrade in the classification to Class B_2 with an MCL/MCLG, respectively, of 0.005 mg/L and zero. A Class C designation was also seriously considered in that proposal. Public comment was solicited and a decision was made based on the total weight of the evidence, that is, one positive mouse study in two mouse species, a partially corroborating study in one species, no human evidence, no replication of the results in animals, negative evidence of carcinogenicity in structurally similar compounds, negative mutagenicity studies, uncertainties with mode of administration, and controversy surrounding the significance of the rat kidney and mouse liver tumor results.

In changing the originally proposed classification of *para*-dichlorobenzene from D to C, the MCL/MCLG of 0.75 mg/L was adjusted to 0.075 mg/L. An additional safety factor of 10 was used when calculating the MCLG for a Class C substance. EPA will continue to reassess this and all other substances as new information becomes available.

OTHER ASPECTS OF THE RULE

In addition to the establishment of MCLGs and MCLs for the eight VOCs, this rule covers the following conditions for regulating those chemicals as specified:

- BAT for treatments under SDWA Sections 1412 and 1415 (variances)
- monitoring requirements and compliance determination
- public notification and reporting requirements
- analytical methods for detection
- laboratory certification criteria
- allowable point-of-entry (POE) and point-of-use (POU) devices and bottled water uses to achieve compliance variances and exemptions of control techniques for VOCs;

and simultaneously specifies monitoring of SDWA Section 1445, the unregulated contaminants.

Public Water Systems

An additional definition was added for public water systems for which directives of this rule apply. Public water systems are divided into community and noncommunity systems. A community water system (CWS) is one which serves at least 15 connections used by year-round residents or regularly serve at least 25 year-round residents. Noncommunity water systems (NCWSs) are, by definition, all other water systems and include transient (e.g., campgrounds or gas stations) and nontransient (e.g., schools, workplaces, or hospitals that have their own water supplies and serve the same population over six months of a year) systems.

Instead of changing the definition of community water system as was proposed in the November 1985 notice, EPA has promulgated a definition of a "nontransient noncommunity water system" (NTNCWS) and applied it to the National Primary Drinking Water Standards (NPDWSs) for the eight VOCs in addition to the already defined systems. According to the EPA, a noncommunity nontransient water system is

> a public water system that is not a regular community water system and that supplies at least 25 of the same people over six months per year.[2]

The purpose of the change was to protect nonresidential populations of more than 25 people who, because of regular long-term exposure, might incur long-term risks of adverse health effects similar to those incurred by those residential populations. The change was designed to include systems serving more than 25 persons in such places as workplaces, offices, and schools that have their own water supplies and where their constituents, while there, might consume from one-third to one-half or more of the normal daily water consumption.

Unregulated Contaminants

Section 1445 of the SDWA, as mentioned earlier, requires that public water systems conduct a monitoring program for unregulated contaminants promulgated December 19, 1987. Each system must monitor at least once every five years for unregulated contaminants unless EPA requires more frequent monitoring. This data will assist EPA in determining whether regulations for these contaminants will be necessary, and if so, what levels might be appropriate.

A list of 51 chemicals (see Table 2) has been chosen and separated into three groups:[3]

Table 2. Unregulated Contaminants under SDWA Section 1445

List 1: Monitoring Required for All Systems	
Bromobenzene	1,1-Dichloroethane
Bromodichloromethane	1,1-Dichloropropene
Bromoform	1,2-Dichloropropane
Bromomethane	1,3-Dichloropropane
Chlorobenzene	1,3-Dichloropropene
Chlorodibromomethane	2,2-Dichloropropane
Chloroethane	Ethylbenzene
Chloroform	Styrene
Chloromethane	1,1,2-Trichloroethane
o-Chlorotoluene	1,1,1,2-Tetrachloroethane
p-Chlorotoluene	1,1,2,2-Tetrachloroethane
Dibromomethane	Tetrachloroethylene
m-Dichlorobenzene	1,2,3-Trichloropropane
o-Dichlorobenzene	Toluene
trans-1,2-Dichloroethylene	*p*-Xylene
cis-1,2-Dichloroethylene	*o*-Xylene
Dichloromethane	*m*-Xylene
List 2: Required for Vulnerable Systems	
Ethylene dibromide (EDB)	
1,2-Dibromo-3-Chloropropane (DBCP)	
List 3: Monitoring Required at the State's Discretion	
Bromochloromethane	*n*-Propylbenzene
n-Butylbenzene	*sec*-Butylbenzene
Dichlorodifluoromethane	*tert*-Butylbenzene
Fluorotrichloromethane	1,2,3-Trichlorobenzene
Hexachlorobutadiene	1,2,4-Trichlorobenzene
Isopropylbenzene	1,2,4-Trimethylbenzene
p-Isopropyltoluene	1,3,5-Trimethylbenzene
Naphthalene	

List 1—Monitoring required for all CWSs and NTNCWSs. Compounds can be readily analyzed.

List 2—Monitoring required only for systems vulnerable to contamination by these compounds. Compounds have limited localized occurrence potential and require some specialized handling.

List 3—The state decides which systems would have to analyze for these contaminants, which include compounds that do not elute within reasonable retention time using packed column [treatment] methods, or are difficult to analyze because of high volatility or instability, and are much less likely to be present in drinking water.

The monitoring methods for the unregulated VOCs are similar to those required for the regulated VOCs, so that public water systems are encouraged to use the same samples for all the analyses and to have the analyses of the unregulated VOCs performed with the analyses for the regulated VOCs, thereby reducing costs for both sampling and analysis.

This list of 51 chemicals also contains some of the disinfection by-products that are scheduled to be promulgated by June 1989 as part of Phase IV of the regulatory program. Other disinfection by-products will be extracted from the Drinking Water Priority List (DWPL).

Along with the VOC rule, two proposals were announced; a list of substitutes on and off of the original list of 83 contaminants and a list of 25 additional substances.[4] The lists of both proposals were added to the DWPL, the final version of which was published in December 1987.

REFERENCES

1. *Federal Register,* Vol. 52, No. 130 (1987).
2. *Federal Register,* Vol. 52, No. 130, p. 25695 (1987).
3. *Federal Register,* Vol. 52, No. 130, p. 25710 (1987).
4. *Federal Register,* Vol. 52, No. 130, p. 25720 (1987).

CHAPTER 5

National Primary Drinking Water Regulations for Synthetic Organic Chemicals, Inorganic Chemicals, and Microbiological Contaminants

Joseph A. Cotruvo and Marlene Regelski

PHASE II: ORGANICS AND INORGANICS

Phase II regulatory goals satisfy the statutory requirements of the Safe Drinking Water Act of 1974 (SDWA) and Amendments to set 40 maximum contaminant level goals (MCLGs) and 40 maximum contaminant levels (MCLs) plus the monitoring of 51 contaminants by June 1988 (see Table 1). Microbials will be handled concurrently with filtration criteria for an expected promulgation date of June 1988.

The Phase II chemicals proposed on November 13, 1985, only included MCLGs. Since then, the SDWA Amendments stipulated that MCLs and MCLGs must be proposed and promulgated simultaneously. Promulgation has been slower for these chemicals because of inadequate data on the occurrence of these substances in environmental drinking water and on the treatment technology. However, there is enough information specified as needed by law to regulate these contaminants.

Occurrence and Health Effects

Thirty synthetic organic chemicals and 10 inorganic chemicals are included with Phase II. These 40 chemicals represent a widely varied

Table 1. Phase II Proposed Chemicals

	Tentative MCLGs for Inorganic Chemicals:	
(1)	Arsenic	zero
(2)	Asbestos	7 million fibers/liter
(3)	Barium	4.7 mg/L
(4)	Cadmium	0.005 mg/L
(5)	Chromium	0.1 mg/L
(6)	Copper	1.3 mg/L
(7)	Mercury	0.004 mg/L
(8)	Nitrate[a]	10 mg/L (as N)
(9)	Nitrite	1 mg/L (as N)
(10)	Selenium	0.05 mg/L
	Tentative MCLGs for Organic Chemicals:	
(1)	Acrylamide	zero
(2)	Alachlor	zero
(3)	Aldicarb	0.01 mg/L
(4)	Aldicarb sulfoxide	0.01 mg/L
(5)	Aldicarb sulfone	0.04 mg/L
(6)	Atrazine	0.003 mg/L
(7)	Carbofuran	0.04 mg/L
(8)	Chlordane	zero
(9)	Dibromochloropropane	zero
(10)	*o*-Dichlorobenzene	0.6 mg/L
(11)	*cis*-1,2-Dichloroethylene	0.07 mg/L
(12)	*trans*-1,2-Dichloroethylene	0.07 mg/L
(13)	1,2-Dichloropropane	zero
(14)	2,4-D	0.07 mg/L
(15)	Epichlorohydrin	zero
(16)	Ethylbenzene	0.7 mg/L
(17)	Ethylene dibromide	zero
(18)	Heptachlor	zero
(19)	Heptachlor epoxide	zero
(20)	Lindane	0.0002 mg/L
(21)	Methoxychlor	0.3 mg/L
(22)	Monochlorobenzene	0.3 mg/L
(23)	PCBs	zero
(24)	Pentachlorophenol	0.2 mg/L
(25)	Styrene	0.1 mg/L
(26)	Tetrachloroethylene	zero
(27)	Toluene	2.0 mg/L
(28)	Toxaphene	zero
(29)	2,4,5-TP	0.05 mg/L
(30)	Xylene	12 mg/L

group of contaminants, each with its unique contamination problem. The synthetic organics may be found where there is manufacturing; pesticides, where there is agricultural development; and the inorganics, both in natural geologic formations and in treatment and conveyance mechanisms for drinking water supplies and sources (i.e., lead).

The health effects produced by these chemicals are as varied as their uses. Some are potent neurotoxins, others are organ-specific toxicants,

Table 1. Continued

Tentative Proposed MCLs for Inorganic Chemicals:		
(1)	Arsenic	0.03 mg/L
(2)	Asbestos	7 million fibers/liter
(3)	Barium	4.7 mg/L
(4)	Cadmium	0.005 mg/L
(5)	Chromium	0.1 mg/L
(6)	Copper	1.3 mg/L
(7)	Mercury	0.004 mg/L
(8)	Nitrate[a]	10 mg/L (as N)
(9)	Nitrite	1 mg/L (as N)
(10)	Selenium	0.05 mg/L
Tentative Proposed MCLs for Organic Chemicals:		
(1)	Acrylamide	0.00065 mg/L
(2)	Alachlor	0.002 mg/L
(3)	Aldicarb	0.01 mg/L
(4)	Aldicarb sulfoxide	0.01 mg/L
(5)	Aldicarb sulfone	0.04 mg/L
(6)	Atrazine	0.003 mg/L
(7)	Carbofuran	0.04 mg/L
(8)	Chlordane	0.002 mg/L
(9)	Dibromochloropropane	0.0002 mg/L
(10)	*o*-Dichlorobenzene	0.6 mg/L
(11)	*cis*-1,2-Dichloroethylene	0.07 mg/L
(12)	*trans*-1,2-Dichloroethylene	0.07 mg/L
(13)	1,2-Dichloropropane	0.005 mg/L
(14)	2,4-D	0.07 mg/L
(15)	Epichlorohydrin	0.003 mg/L
(16)	Ethylbenzene	0.7 mg/L
(17)	Ethylene dibromide	0.00005 mg/L
(18)	Heptachlor	0.0004 mg/L
(19)	Heptachlor epoxide	0.0002 mg/L
(20)	Lindane	0.0002 mg/L
(21)	Methoxychlor	0.3 mg/L
(22)	Monochlorobenzene	0.3 mg/L
(23)	PCBs	0.0005 mg/L
(24)	Pentachlorophenol	0.2 mg/L
(25)	Styrene	0.1 mg/L
(26)	Tetrachloroethylene	0.005 mg/L
(27)	Toluene	2.0 mg/L
(28)	Toxaphene	0.005 mg/L
(29)	2,4,5-TP	0.05 mg/L
(30)	Xylene	12 mg/L

Note: These lists of maximum contaminant levels (MCLs) and maximum contaminant level goals (MCLGs) are all tentative, having not been formally proposed by EPA as of this writing.

[a] MCLG for *total* nitrate and nitrite = 10 mg/L (as N).

and some are apparent or known carcinogens. Because of this, the approach to setting MCLGs and MCLs for each chemical must be very comprehensive.

Organics

Over half of the organics are pesticides, which are being more frequently detected in drinking water. Unlike other synthetic organics used in manufacturing products and as additives, pesticides are manufactured to be toxic and are applied directly to the ground to kill pests. Some are herbicides registered for aquatic application and are either applied directly to water or migrate to drinking water sources from runoff. Their widespread use and direct access to water supplies make some of them of special concern for drinking water contamination.

Aldicarb. Aldicarb, a pesticide widely used to control mites and nematodes, is applied directly to plants and into the soil where it migrates readily to groundwater. It rapidly degrades in water to its two metabolite forms of aldicarb sulfoxide and sulfone, sulfoxide being the more toxic. Aldicarb and its metabolites are potent, fast-acting cholinesterase enzyme inhibitors that cause symptoms such as weakness, gastric cramps, nausea, salivation, sweating, and vomiting. Aldicarb has not been shown to be carcinogenic and its toxic effects in low doses seem to be reversible because of the human body's spontaneous recovery of cholinesterase production. The proposed MCL and MCLG of 0.010 mg/L is based on the acute toxicity of the sulfoxide metabolite. An MCL and MCLG of 0.04 mg/L is proposed for the sulfone metabolite found alone.

Ethylene dibromide. Ethylene dibromide (EDB) is another pesticide whose application has been publicized, because of its use as a grain fumigant. It has been detected in drinking water in several states, including California, South Carolina, Washington, Connecticut, Massachusetts, Georgia, and Florida, at parts-per-billion levels. EDB was detected in up to 30% in raw grain and 8% in grain products at levels as high as 5400 parts per billion. In addition to its pesticide use, it is also added to leaded gasoline as a lead scavenger.

Health effects associated with short exposures to EDB have included lung, liver, spleen, kidney, and central nervous system (CNS) toxicity. Repeated exposure produced liver, stomach, adrenal cortex, and reproductive system toxicity. It is mutagenic and a potent carcinogen (Class B_2) in rats and mice. EDB's proposed MCLG is zero.

Epichlorohydrin. Epichlorohydrin, another organic chemical, not a pesticide, produces a wide range of health effects in the form of acute toxicity, mutagenicity, and carcinogenicity (Class B_2). This compound is a component of epoxy resins sometimes used to coat the inside of water tanks and pipes and is a raw material for epoxy and phenoxy resins and flocculants.

In addition to causing acute respiratory suppression through inhalation at high levels, epichlorohydrin is dermally and orally toxic, causing extensive irritation at the point of exposure as well as kidney and liver systemic toxicity. Epichlorohydrin is a contaminant of polymers used in the clarification and storage of potable water and in food processing. It has been detected in waste and is considered mobile in water. It is being regulated in drinking water because of its possible carcinogenic risks.

Inorganics

Inorganic chemicals are mostly naturally occurring contaminants prevalent in natural geological formations. Some are also consistently found in drinking water supplies from man-made sources, i.e., copper, lead, chromium, asbestos pipes, and plumbing supplies. These metals leach into water sources either naturally or as a result of corrosion of the pipes and plumbing supplies.

Lead. Lead, an inorganic metal ubiquitous in water, also is of great concern. Lead contamination of drinking water is mostly due to corrosion of lead pipe, solder, and flux in the public water supply system after treatment.

Asbestos. Asbestos poses a different problem for drinking water regulators due to its fibrous nature. Asbestos is naturally occurring, but also is used in drinking water supply pipes and can be leached into drinking water supplies by corrosive water. The health concern with asbestos is through inhalation exposure. Human exposure to asbestos in drinking water occurs primarily via ingestion, but exposure via inhalation can occur as a result of the use of humidifiers and, possibly, showers. The U.S. Environmental Protection Agency (EPA) is proposing an MCLG and an MCL for asbestos of 7 million fibers per liter of fibers greater than 10 μm in length.

Arsenic. Arsenic constitutes an interesting case for regulation because it has been considered an essential nutritional dietary element by some scientists. It is also a known carcinogen, causing skin cancer by oral exposure in some cases. Options being considered are to set the MCLG at zero as for all carcinogens or to set the MCLG at some finite number, considering that skin cancer is treatable. Occurrence in water can result from both natural processes and industrial activities, including smelting operations, use of arsenical pesticides, and industrial waste disposal.

Arsenic, lead, and copper have recently been shifted from the Phase II group to other regulations. Lead and copper have been proposed as part of a regulation to install corrosion control to reduce leaching from piping solder and water taps.

Anticipated Regulatory Changes

There are quite a few suggested changes to the proposed rule announced November 13, 1985. Four of the chemicals have had MCLs rounded off from the proposed rule, and six MCLs were reproposed based on public comments and new scientific information. Those which may be changed include arsenic, lead, mercury, aldicarb, 1,2-dichloropropane, monochlorobenzene, and xylene. Tetrachloroethylene was originally set to be proposed with the eight volatile organic chemicals (VOCs), but issues were not settled at the time of final notice. During the next public comment period, they will again be evaluated before the final notice is published.

Secondary MCLs are set as aesthetic drinking water standards to regulate color, odor, and taste. Under Phase II, the secondary MCLs are being proposed for the following chemicals: aluminum, *o*-and *p*-dichlorobenzene, 1,2-dichloropropane, silver, styrene, toluene, and xylene.

For all synthetic organics, granular activated carbon (GAC) will be specified as best available technology (BAT). For DBCP, 1,2-dichloropropane, *cis*-1,2-dichloroethylene, *trans*-1,2-dichloroethylene, *o*-dichlorobenzene, ethylene dibromide, ethylbenzene, monochlorobenzene, styrene, toluene, and xylene, packed tower aeration (PTA) will be specified in addition. For inorganics, probable BAT treatment techniques are shown in Table 2.The same BAT for variances under inorganics are specified. Monitoring and reporting requirements and compliance determination, analytical methods of detection, lab certification criteria, monitoring for unregulated contaminants, and regulatory impact analysis will also be stipulated.

MICROBIALS AND SURFACE WATER TREATMENT

Originally part of Phase II, microbiological contaminant regulations have been separated and are now merged with surface water treatment/filtration regulations. The time schedules for promulgation of the final rules for surface water treatment and coliform bacteria are currently the same. The proposed rules were announced on November 3, 1987 (52 FR 42179; 52 FR 42224) and May 1988 (53 FR 16348).

Table 2. BAT Under Section 1412 of SDWA for Inorganic Chemicals

Chemical	Technology
Asbestos	Coagulation/Filtration Corrosion control
Arsenic	Activated alumina Ion exchange Lime softening Reverse osmosis Coagulation/Filtration
Barium	Ion exchange Lime softening Reverse osmosis
Cadmium	Ion exchange Reverse osmosis Coagulation/Filtration Corrosion control Lime softening
Chromium	Coagulation/Filtration Ion exchange Lime softening Reverse osmosis
Copper	Ion exchange Reverse osmosis Coagulation/Filtration Lime softening Corrosion control
Mercury	Granular activated carbon Coagulation/Filtration + powdered activated carbon Lime softening Reverse osmosis
Nitrate/Nitrite	Ion exchange Reverse osmosis Oxidation (nitrite)
Selenium	Activated alumina Lime softening Coagulation/Filtration Reverse osmosis

Note: BAT = best available technology.

These treatment rules will standardize and upgrade both monitoring and treatment processes and disinfection standards. Local water suppliers throughout the United States will now be directed by EPA to filter their water and/or to disinfect it under certain specified conditions to protect against coliform bacteria, *Giardia lamblia,* heterotrophic bacteria, *Legionella hemophila,* turbidity, and viruses. Most people do not realize that many water supply systems do not filter or do not disinfect. Of the 9800 drinking water systems in the United States using surface water, 3000 systems, serving approximately 21 million people, currently do not filter. EPA's current standards, in effect since 1977, protect for coliform bacteria and turbidity.

Although our drinking water in the United States is among the safest in the world, thousands of cases of waterborne disease occur every year. While treatment may be adequate at the drinking water source, conditions may exist in the distribution system that permit regrowth of microbial, bacterial, and viral contaminants. In recent years, outbreaks of legionnaires' disease and giardiasis (also called backpacker's disease) have brought waterborne disease more into the public eye.

Types of Microbial Contaminants

Coliform bacteria come from human and animal waste. While common in the environment and generally not themselves harmful, their presence indicates that the water may be contaminated with disease-causing organisms. Total coliform bacteria regulations apply to all 200,000 public groundwater and surface water systems, both community and noncommunity supplies.

Giardia lamblia is a species of protozoa that originates in human and animal waste. Giardiasis, the disease it causes, has symptoms that are flu-like but usually more severe, causing diarrhea, nausea, and dehydration that can last for months in some cases. Backpackers who drink unfiltered, nondisinfected water from mountain streams often come down with this, thus the name "backpacker's disease." *Giardia* cysts can be filtered out of water because of their size or can be inactivated by a rigorous disinfection process.

Heterotrophic bacteria are bacteria that use only organic materials as their food source. Turbidity is a measure of cloudiness or clarity of water, which is indicative of excess organic material (including animal or human waste) in water. Therefore, measuring for these two items can give the water suppliers an indication when disease-causing microorganisms may be present or information on the effectiveness of treatment processes.

Legionella hemophila is a species of bacteria that causes severe pneumonia-like symptoms (i.e., legionnaires' disease), especially in a weaker population such as the elderly.

Viruses, enteric and other, cause diseases such as hepatitis A and gastroenteritis.

Regulation Specifics

These proposed surface water treatment/filtration regulations set criteria for the states to determine which water systems will have to install filtration or update existing filtration facilities and/or disinfect:

- All surface water systems will now have to disinfect.
- All surface water systems must filter unless they meet source water quality criteria and site specific conditions.
- Only qualified operators will be entitled to operate the systems. All systems will need to achieve at least 99.9% removal and/or inactivation of *Giardia* cysts and 99.99% removal and/or inactivation of enteric viruses.

Filtration is not required if:

- A system meets source water quality criteria (coliform and turbidity levels);
- A system meets the following site specific conditions:
  1. has disinfection which achieves 99.9% and 99.99% inactivation of *Giardia* cysts and viruses, respectively
  2. maintains watershed control/satisfies sanitary survey requirements
  3. has no history of waterborne disease outbreak without making treatment corrections
  4. complies with long-term coliform MCL
  5. meets with total trihalomethanes (TTHMs) MCL

Finally, local water system operators must report to their state governments monthly on their progress in meeting federal rules and within 48 hours on waterborne disease outbreaks, and operators of both filtered and unfiltered water systems must meet federal requirements within four years after the final rule is issued.

These rules, following modifications derived from public comments, are expected to be promulgated by mid-1989.

CHAPTER 6

National Primary Drinking Water Regulations for Radionuclides

Edward V. Ohanian

INTRODUCTION

In 1976, the U.S. Environmental Protection Agency (EPA) promulgated the National Interim Primary Drinking Water Regulations (NIPDWR). The Safe Drinking Water Act (SDWA) of 1974, as amended in 1986, requires that the EPA now promulgate the National Revised Primary Drinking Water Regulations (NRPDWR). Revised regulations are developed in two steps by establishing maximum contaminant level goals (MCLGs) and maximum contaminant levels (MCLs). Under the 1986 Amendments to the SDWA, these two values are published concurrently for each contaminant—first as proposed regulations and then as final regulations.

The MCLGs are nonenforceable health goals set at levels that should result in no known or anticipated adverse health effects and allow an adequate margin of safety. The MCLs are the enforceable standards and must be set as close to the MCLGs as feasible. *Feasible* means "with the use of the best technology, treatment techniques and other means, which the Administrator finds are generally available (taking cost into consideration)." The selection of contaminants for regulation depends on the availability of information on occurrence, health effects, analytical methodology, treatment, and cost.

The NIPDWRs include both natural and man-made radionuclides. The standards for natural radionuclides include a gross alpha particle standard of 15 pCi/L and a combined radium-226 and radium-228 standard of 5 pCi/L. Both radon and uranium are specifically excluded from the gross alpha particle activity because of lack of data and information concerning their occurrence and toxicity. The interim standard for man-made radionuclides is a total dose equivalent of 4 millirems per year.

The MCLGs under development for radionuclides include values for radium-226 and radium-228 separately, natural uranium, radon-222, gross alpha particle activity and man-made radionuclides. All are estimated to pose carcinogenic risks to humans. The House Report on the 1974 legislation states that when there is no threshold in the dose-response curve for a drinking water contaminant (e.g., a carcinogen), the MCLGs are required to be set at zero. This is because the MCLG is to be set at a level for which there are no known or anticipated adverse health effects.

This chapter provides a summary of the information compiled to date by the EPA toward development of the NRPDWRs for radionuclides. Emphasis is given to information available for radium, uranium, and radon, as these are the radioisotopes of primary concern with regard to current regulatory efforts.

OCCURRENCE

Radium-226

Available occurrence data for radium-226 was estimated using compliance monitoring data developed under the interim regulations, an EPA survey of 2500 public water systems and limited studies of east coast and midwest aquifers. In 1980 and 1981, EPA's Office of Radiation Programs conducted a nationwide survey of 2500 public water systems involving 27 states and representing 45% of the drinking water consumed in the United States. Samples were taken primarily from groundwater systems serving more than 1000 people. A gross alpha particle screening step of 5 pCi/L was employed.

The average, population-weighted concentration is shown in Table 1. The estimated average concentration for all drinking water supplies, both surface and groundwater, ranged from 0.3 to 0.8 pCi/L. Actual concentrations for radium-226 fall in the range from the detection limit to as high as 200 pCi/L. Between 300 and 3000 public drinking water

Table 1. Population-Weighted Average Concentration of Natural Radionuclides in U.S. Community Drinking Water

Radionuclide	Concentrations[a] (pCi/L)
Radium-226	0.3–0.8
Radium-228	0.4–1.0
Uranium-natural	0.3–2.0
Radon-222	50–300
Lead-210	<0.11
Polonium-210	<0.13
Thorium-230	<0.04
Thorium-232	<0.01

[a] Average of surface and groundwater supplies.

supplies are estimated to have a radium-226 concentration exceeding 1 pCi/L, a level corresponding to projected excess cancer risks of 1 in 100,000 persons.

Radium-228

Available data on the occurrence of radium-228 in groundwater are limited; however, several studies have shown that natural waters have approximately equal activities for radium-226 and radium-228. Using this information along with other available information on radium-228, the average population-weighted concentration in public drinking water supplies, as indicated in Table 1, is estimated to range from 0.4 to 1.0 pCi/L.

Analysis of the geochemical transport properties of radium-228 indicated that aquifers with high levels of radium-228 would be expected to be granite, arcosic sand, and quartose sandstone aquifers with high total dissolved solids. Aquifers having low activity of radium-228 would be those whose geology is either carbonate metamorphic rock or quartose sand or sandstone and basic igneous rocks.

Natural Uranium

Natural uranium consists of three isotopes: uranium-234, uranium-235, and uranium-238. These isotopes are found in both surface and groundwater supplies with relative occurrence rates of 0.006%, 0.72%, and 99.27%, respectively. The concentrations of natural uranium in surface and groundwaters have been estimated using the National Uranium Resource Evaluation (NURE) study conducted by the U.S. Geological Survey (USGS) in the latter part of the 1970s. Over 34,000 surface water and more than 55,000 groundwater samples were evaluated by delayed neutron activation analysis. Natural uranium concentrations in groundwater were found to be generally higher than in surface water. Approximately 28,000 of the samples were identified as likely to have come from public drinking water sources. As seen in Table 1, the population-weighted average concentration of uranium in drinking water ranged from 0.3 to 2.0 pCi/L. Actual measurements ranged from the detection limit up to approximately 600 pCi/L, with only a few supplies exceeding the level of 50 pCi/L. The arithmetic average for the ground and surface-water samples was 2 pCi/L. Between 100 and 2000 public drinking water supplies are estimated to have concentrations exceeding 7 pCi/L, a level corresponding to a projected cancer risk of one in 100,000.

Radon

Radon concentrations were measured by EPA's Office of Radiation Programs for approximately 2500 larger water systems (greater than 1000 people) in the United States. Roughly 1500 more measurements existed from various independent sources. As seen in Table 1, the average population-weighted concentration from all drinking water supplies in the United States ranged from 50 to 300 pCi/L. Actual concentrations were measured from the detection limit to as high as 500,000 pCi/L. As expected, virtually all of the radon in drinking water was found in groundwater supplies. Available data, while limited, confirm that surface water has radon-222 concentrations generally less than the detectable level (5 to 10 pCi/L).

Other Radionuclides

Radionuclides other than radium, uranium, and radon, as listed in Table 1, are shown in the form of the upper limit because they have not been detected in drinking water, due to their insolubility in water. Thus, these upper limits are estimates based on their chemical and physical properties.

RELATIVE SOURCE CONTRIBUTIONS

Table 2 shows the average relative source contributions to the daily human intake of naturally occurring radionuclides. As seen in this table, the contribution of radium-226 from food is approximately the same as that from drinking water, averaging 1.35 pCi/day. Similar daily intakes are shown for radium-228. For uranium, the daily intake from food is about 1 pCi/day, while drinking water contributes from 0.6 to 4 pCi/day. For radon, indoor air is clearly the predominant exposure source.

Table 2. Average Relative Source Contribution to the Daily Intake of Natural Radionuclides.

Radionuclide	Source	pCi/day
Radium-226	Air	0.007
	Food	1.1–1.7
	Drinking water	Generally small from surface supplies, 0.6–2
Radium-228	Air	0.007
	Food	1.1
	Drinking water	0.6–2
Uranium-234 and Uranium-236	Air	0.0007
	Food	0.37–0.9
	Drinking water	0.6–4
Lead-210	Air	0.3
	Food	1.2–3
	Drinking water	<0.02
Polonium-210	Air	0.06
	Food	1.2–3
	Drinking water	<0.02
Radon-222	Outdoors	970
	Indoors	8100
	Drinking water	100–800
Thorium-230	Air	0.0007
	Food	Probably negligible
	Drinking water	<0.06
Thorium-232	Air	0.0007
	Food	Negligible
	Drinking water	<0.02

The primary source of indoor air radon is soil. It is estimated that drinking water contributes 1–7% of total indoor air radon levels.

These drinking water contributions to indoor air radon levels derive from showers, baths, dishwashers, clothes washers, toilets, and other related sources.

HEALTH EFFECTS

Radium

Radium has an affinity for skeletal tissue and is known to induce both bone sarcomas and head carcinomas. Health effects of radium were reported in an epidemiology study conducted among 3700 persons in the United States exposed to radium-226 and radium-228 during the process of painting watch dials, medical administration procedures, or by other means. A total of 85 cases of bone sarcoma and 36 cases of head carcinoma were observed. The results of this study are consistent with the dosimetric model predictions of dose and risk in the International Commission on Radiological Protection (ICRP) Report 30.[1] These predictions are being used to develop the proposed MCLG.

Natural Uranium

Uranium also has an affinity for skeletal tissue; however, there are no epidemiology studies demonstrating radiotoxicity. It is known that between 1% and 5% of ingested uranium deposits in the bone in a manner similar to radium. Additionally, the alpha particle decay of uranium occurs in a manner similar to that of radium. Therefore, dose and risk to bone can be estimated using the same ICRP 30 dosimetric model[1] and the risk factors derived from the BEIR III report of the National Academy of Sciences.[2]

As evidenced by over a century of data from both medical administration to humans as well as from numerous animal studies, the primary chemical toxic effect of natural uranium is on the kidneys. Nephritis (inflammation of the kidneys) and polyuria (excessive urine excretion) are clear symptoms. It is estimated by EPA that health effects to the kidney are of the same order of magnitude as radiotoxic effects to bone.

Radon

Health effects of inhaled radon in humans has been shown in epidemiology studies involving hardrock miners from several countries, including the United States, Canada, Sweden, and Czechoslovakia. From these studies, dose-response data have been produced indicating that the inhaled radon led to lung cancer. The data have been summarized in ICRP Report 32[3] and the dose and risk estimates from that report are used as the basis of EPA's estimates of dose and risk. It is estimated that the health risk due to inhaling radon from drinking water sources is considerably larger than the risk from ingesting drinking water containing radon.

RISK ESTIMATES

Using the population-weighted concentrations from Table 1 and the individual estimates produced from the ICRP 30 dosimetric model[1] and the BEIR III risk weighting factors,[2] the population risks for radionuclides in drinking water have been estimated. These values are summarized in Table 3. The population risk estimate for radon dominates, being somewhere in the range of 2000 to 40,000 excess lung cancers. To estimate the population risk due to man-made radionuclides, the radio-

Table 3. Estimates of Population Risk for Some Radionuclides in Drinking Water[a]

Radionuclide	Estimates of Lifetime[b] Population Risk (number of fatal cancers due to current exposures in drinking water)
Radium-226	200–4,000
Radium-228	200–4,000
Uranium-natural	40–1,000
Radon-222	2,000–40,000
Strontium-90[c]	5–160

[a] Rounded off to one significant figure.

[b] 70 years.

[c] Surrogate for all man-made radionuclides.

nuclide strontium-90 was selected. It has been estimated that this radionuclide supplies the largest contribution to the population risk for all man-made radionuclides. The resulting estimate of lifetime risk due to man-made radionuclides in drinking water is in the range of 5 to 160 excess cancers.

Tables 4, 5, and 6 provide estimates of the number of public drinking water supplies that exceed various levels for radium-226, uranium, and radon. As expected, the number of drinking water supplies and populations that exceed various levels of radon increase as the risk level decreases.

The occurrence and risk levels for radioactivity in drinking water are summarized in Table 7. The existing standard for combined radium-226 and radium-228 is 5 pCi/L, which falls somewhere in the middle of the table, reading from top to bottom. The risk level for ingesting 2 liters per day of drinking water containing 5 pCi/L of radium-226 is in the order

Table 4. Estimates of the Number of Public Drinking Water Supplies That Exceed Various Levels of Radium-226[a]

Lifetime Risk Level	Radium-226 Concentration (pCi/L)	Annual Effective Dose Equivalent (mrem/yr)	Number of Public Drinking Water Supplies That Exceed the Concentration in Column 2
10^{-3}	100	100	1–10
10^{-4}	10	10	30–300
10^{-5}	1	1	300–3000
10^{-6}	0.1	0.1	b

[a] Rounded off to one significant figure.

[b] Below detection level.

Table 5. Estimates of the Number of Public Drinking Water Supplies That Exceed Various Levels of Natural Uranium[a]

Lifetime Risk Level	Uranium Concentration (pCi/L)	Annual Effective Dose Equivalent (mrem/yr)	Number of Public Drinking Water Supplies That Exceed the Concentration in Column 2
10^{-3}	700	100	1–10
10^{-4}	70	10	20–500
10^{-5}	7	1	100–2000

[a] Rounded off to one significant figure.

of 4×10^{-5} cancers per lifetime. A corresponding concentration of uranium would be 40 pCi/L, and a corresponding level for radon would be a few hundred pCi/L. From the data on the occurrence of radon, it can be seen that a few hundred pCi/L is about the average level for radon. That is, if the standard would be set for radon at the same risk level as the existing radium-226 standard, roughly half of the groundwater supplies in the United States would exceed the standard.

ANALYTICAL METHODS

Table 8 lists the analytical methods for determining concentrations of radionuclides in drinking water. Many methods are available for analysis of radium; however, several of these methods have not yet been validated. The coincidence and gamma ray spectroscopy methods may be validated in the near future. The gross alpha particle activity methods for dealing with high solids have been validated. The uranium laser-induced fluorometric method will be validated in the future. The three methods

Table 6. Estimates of the Number of Public Drinking Water Supplies That Exceed Various Levels of Radon[a]

		Number That Exceed the Concentration in Column 2	
Lifetime Risk Level	Concentration (pCi/L)	Public Drinking Water Supplies	Population (thousands)
10^{-3}	10,000	500–4000	20–300
10^{-4}	1,000	1,000–10,000	200–4,000
10^{-5}	100	5,000–30,000	10,000–100,000
10^{-6}	10	10,000–40,000	50,000–100,000

[a] Rounded off to one significant figure.

Table 7. Summary of Occurrence and Risk Levels for Radioactivity in Drinking Water[a]

	Annual Effective Dose Equivalent[b] (mrem/year)	pCi/L			
		Radium-226	Radium-228	U_{nat}[c]	Radon-222
Estimated lifetime risk level					
10^{-3}	100	100	200	700	10,000
10^{-4}	10	10	20	70	1,000
10^{-5}	1	1	2	7	100
10^{-6}	0.1	0.1	0.2	0.7	10
Concentration averages, population-weighted					
All supplies		0.3–0.8	0.4–1.0	0.3–2.0	50–300
Groundwater supplies		1.6	1.8	3	approx. 400
Surface water supplies		—	—	1	—
Actual concentration		0–200	0–50	0–600	0–500,000

[a] The calculations in this table involve uncertainties of the order of 4 to 5.

[b] Rounded off to one significant figure. Note that the dose limit for man-made radioactivity in drinking water under the interim regulations is 4 mrem/year, at the end of 70 years.

[c] Using $f_1 = 0.05$.

listed for radon, namely, liquid scintillation, solid state nuclear track detector, and Lucas cell methods, are currently being validated.

TREATMENT TECHNOLOGY

Several methods are available in the literature for treating drinking water supplies for radium. Recent studies indicate that the techniques available to remove uranium from drinking water supplies are anion exchange, lime softening at high pH, and reverse osmosis. The two available methods for removing radon from drinking water are granular activated carbon and aeration.

The EPA has developed preliminary cost estimates for technologies that may feasibly remove radionuclides from drinking water. Depending on the amount of water treated, estimated costs range from 30 to 80 cents per 1000 gallons for cation ion exchange, 30 to 110 cents per 1000 gallons for iron and manganese treatment, and 160 to 320 cents per 1000 gallons for reverse osmosis.

Preliminary cost estimates for aeration technique range from 10 to 75 cents per 1000 gallons for systems serving about 100,000 people and 100 to 500 people, respectively.

Table 8. Analytical Methods for Radionuclides

Method	Validation[a]
Radium	
Alpha-emitting radium isotopes (method 903.0)	Yes
Radium-226-radon emanation techniques (method 903.1)	Yes
New York State Department of Health (radium-226 and -228)	No
Total radium (method 304)	Yes
Radium-226 (method 305)	Yes
Coincidence spectrometry	No
Gamma ray spectrometry (radium-226 and -228)	No
Solid state nuclear track detector	Terradex is preparing equivalency test results
Radiochemical determination of radium-226 in water samples (method Ra-03)	No
Radiochemical determination of radium-228 in water samples (method Ra-05)	No
Radium-228 by liquid scintillation counting (method 904.1)	No
Radium-228 (method 904.0)	No
Gross Alpha Particle Activity	
Gross alpha and gross beta radioactivity (method 900.0)	Yes
Gross radium alpha screening (method 900.1)	Single lab tested
Gross alpha activity in drinking water by coprecipitation (method 00–02)	Single lab tested (being collaboratively tested)
Gross alpha and beta (method 703)	Yes
Gross alpha particle activity (method D-1943)	Yes
Gross Beta Particle Activity	
Gross alpha and beta radioactivity (method 900.0)	Yes
Gross beta particle activity (method D-1890)	Yes
Uranium	
Radiochemical (method 906.0)	Yes
Fluorometric (method 966.1)	Yes
Laser induced fluorometry (method 906.2)	No
ASTM method D-2907	Yes
Radon	
Liquid scintillation (including modification using mineral oil so sample can be mailed)	Underway
Solid state nuclear track detector	Underway
Lucas cell	Underway
Man-Made Radionuclides	
Radioactive cesium (method 901.0)	Single lab tested
Gamma-emitting radionuclides (method 901.1)	Yes
Radioactive iodine (method 902.0)	Yes
Radioactive strontium (method 905.0)	Yes
Tritium (method 906.0)	Yes
Strontium-89, -90 (method 303)	Yes
Tritium (method 306)	Yes
Gamma ray spectroscopy (method D-2459)	Yes

[a] Yes = multi-lab validation.

Preliminary cost estimates for removing radon from household drinking water systems by point-of-entry treatment devices are a capital cost of $400 to $800 for granular activated carbon and about $900 for aeration. Operating costs are estimated at $20 per year and $80 per year for activated carbon and aeration, respectively.

ANTICIPATED REGULATORY ACTIVITIES

In accordance with the regulatory schedule specified by the 1986 Amendments to the SDWA, publication of the final MCLGs and MCLs for radionuclides is anticipated by June 1989.

REFERENCES

1. "Limits for Intakes of Radionuclides by Workers," International Commission on Radiological Protection (ICRP). ICRP Publication 30 (1979).
2. "The Effects on Populations of Exposure to Low Levels of Ionizing Radiation," National Academy of Sciences (NAS), Report of the Advisory Committee on Biological Effects of Ionizing Radiation (BEIR III) (1980).
3. "Limits for Inhalation of Radon Daughters by Workers," International Commission on Radiological Protection (ICRP), ICRP Publication 32 (1981).

CHAPTER 7

Issues in Developing National Primary Drinking Water Regulations for Disinfection and Disinfection By-Products

Joseph A. Cotruvo and Marlene Regelski

INTRODUCTION

The regulatory program of the U.S. Environmental Protection Agency (EPA) for disinfection is being mounted in a concerted manner. The interrelationships between the beneficial uses of disinfectants and the potential health concerns stemming from disinfection by-product residues are being considered simultaneously as the regulatory policy is being developed. For example, the Safe Drinking Water Act of 1974 requires that disinfection be specified as a treatment technology for use by all public water systems. Therefore, EPA must determine the appropriate conditions under which disinfection may be used and at which residue concentrations do not adversely affect public health. This must be done via establishment of maximum contaminant levels (MCLs) or other means.

The process of disinfection is used by public water systems in the United States to control pathogenic microorganisms and thus reduce the risk of waterborne disease. The introduction of disinfectants into the water supply, however, has resulted in the presence of undesirable by-products whose toxic properties have resulted in other possible health risks.

Because disinfectants are chemically very reactive substances, they quickly react with the many organic compounds that occur in water. Each reacts individually and can exist in different forms depending upon dosages, pH, temperature, amount of organic substances in the water, and oxidation-reduction processes that might have occurred.

More generally, disinfectants could be termed oxidants, because they oxidize substances in water, i.e., nitrite. Besides their disinfection function, they also assist in flock formation and removal of color from the water.

The pH of the water, which may be regulated to control corrosivity, significantly affects the potency of some disinfectants. All of these competing considerations and more are involved in the analyses that are underway.

Proposed disinfection treatment requirements and by-product regulations are scheduled for proposal in 1989 and promulgation in 1991. There is much to be done before the database for those comprehensive regulations will be sufficient to proceed to optimize drinking water quality. There is just not enough information yet available on the toxicology and health risks to set maximum contaminant level goals (MCLGs) and MCLs in many cases.

Trihalomethanes (THMs), one family of the disinfection by-products, are currently regulated. These compounds are formed in the drinking water during the reaction between chlorine, an effective and widely used disinfectant, and organic matter already in the water. In order to reduce formation of THMs during water treatment, alternative disinfectants are being used to replace free chlorine in many cases. The use of these alternate disinfectants, however, may produce other by-products that may themselves be toxic under some conditions. Table 1 indicates those drinking water disinfectants and disinfection by-products for which development of MCLGs and MCLs are being considered.

The following is a brief summary of the health effects and issues of concern for each disinfectant and disinfection by-product category being considered for regulation.

DISINFECTANTS

Chlorine

Chlorine is the most widely used disinfectant in the United States and has been for more than 60 years.[1] Despite its long and widespread use,

Table 1: Disinfectants and Disinfection By-Products Considered for Development of MCLGs and MCLs

Disinfectants	
Chlorine	Potassium permanganate
Chlorine dioxide	High pH
Chloramine	Ionizing radiation
Ozone	Silver
Iodine	UV light
Bromine	Ferrate
Disinfection By-Products	
Trihalomethanes:	Chlorite and chlorate
Chloroform	Chlorophenols
Bromoform	Chloropicrin
Bromodichloromethane	Cyanogen chloride
Dibromochloromethane	Haloacetonitriles
Chlorinated acetic acids	Iodate
Chlorinated alcohols	Iodide
Chlorinated aldehydes	Bromate
Chlorinated ketones	Bromide

Note: MCL = maximum contaminant level; MCLG = maximum contaminant level goal.

very little information exists on the low-level health effects of ingested chlorine; most laboratory studies have used inhalation as the route of exposure.

The acute toxicity of chlorine in amounts found in drinking water appears to be relatively low. In short-term chlorine toxicity studies with animals, the liver and kidney appear to be the target organs. Chlorine toxicity effects were more pronounced in animals fed high-cholesterol diets. Studies are currently being conducted in monkeys and humans to evaluate whether there are any cardiovascular effects of chlorine ingestion over an extended period of time.[2–5]

No clinical effects were reported in human volunteers consuming chlorine at up to 24 mg/L for 18 days.[6] Chlorinated water has been correlated to bladder and colon cancers in humans in some studies.[7]

Chlorine toxicity via inhalation exposure has been studied in animals and in humans exposed occupationally or through industrial accidents. It is a recognized primary irritant to the mucous membranes of the eyes, nose, and throat, and to the linings of the entire respiratory tract. Almost two thousand casualties during World War I were attributed to gassing by chlorine.[1] The lowest toxic concentration (TC_{LO}) producing respiratory stress in humans is reported to be 15 ppm and the lowest lethal concentration (LC_{LO}) is reported as 430 ppm for a 30-minute exposure.[8]

Additional chronic data and resolution of the issues concerning chlorine's carcinogenicity or cardiovascular toxicity are needed before an MCLG can be determined.

Chlorine Dioxide, Chlorite, and Chlorate

Chlorine dioxide (ClO_2) has often been used in conjunction with chlorine during drinking water disinfection to control chlorophenolic tastes and odors.[2] It was first used in the United States during World War II when chlorine was in short supply. Although more expensive to use than chlorine, chlorine dioxide is a good oxidizing agent and does not produce much in the way of chlorine by-products, but does produce milder oxidation products such as aldehydes. The by-products in which we are mainly interested are chlorate and chlorite.

Chlorine dioxide has also proven to be an effective disinfectant, with nearly 2.5 times the oxidizing power of chlorine. Chlorine dioxide degrades into chlorite (ClO_2^-) and, to a lesser extent, chlorate (ClO_3) during these processes.

The health effects of chlorine dioxide and its conversion products are primarily hematological, presumably due to its oxidizing nature. Human volunteers receiving chlorine dioxide at 0.3 mg/kg/day for 1 day or 0.04 mg/kg/day for 84 days did not show any adverse changes in hematologic values, serum chemistry, or urinalysis parameters.[6]

In various developmental studies conducted with rats, a decreased number of implants and live fetuses, delayed neurodevelopment, and a decreased number of neurons in the cerebellum and forebrain have been reported in the pups of dams given chlorine dioxide at 10 mg/kg/day or more during gestation through weaning.[9-11] The neurotoxic effects observed in rat pups were still evident in 50- to 60-day-old animals whose exposure to chlorine dioxide had been terminated at weaning.[10]

Additional effects observed with exposure to chlorine dioxide include osmotic fragility in red blood cells, decreased DNA synthesis in testes and kidney, and inconsistent changes in glutathione and catalase levels in rats.[12,13]

With exposure to chlorite and chlorate, consistent hematological effects such as decreased red blood cell count, decreased glutathione levels, increased methemoglobin, and decreased hematocrit have been reported in monkeys, mice, rats, and cats exposed to chlorite at 3 mg/kg/day and chlorate at 1 mg/kg/day for up to 84 days in their drinking water.[13-15] The National Academy of Sciences (NAS) has calculated a suggested no adverse response level (SNARL) of 0.3 mg/L for chlorine dioxide and 0.02 mg/L chlorite and chlorate, assuming a 20% contribution from drinking water.[16] However, since these substances are almost unique to drinking water, the assumption of a 20% relative source contribution is probably a significant underestimate. The actual contribution would be closer to at least 90%, yielding a SNARL that could be nearly 5 times the current SNARL estimate.

Chloramine

Chloramine is also used as an alternative to chlorine for disinfection of drinking water. Chloramines are less reactive than chlorine and have their best use as a secondary residual maintenance disinfectant as opposed to a primary pathogenic control agent, because it is more persistent than chlorine. They are not as good in eradicating resistant organisms such as viruses and *Giardia* cysts, but are a cheap way of quenching the formation of halomethanes and other by-products. Chloramines also reduce unaesthetic tastes and odors connected with the formation of chlorophenolic compounds.

The primary toxic endpoint of chloramine in reported studies appears to be hematological effects. Glutathione levels in rats have been reported to increase initially, then decrease significantly with prolonged exposure.[17] Accompanying this effect is increased osmotic fragility and decreased red blood cell count.

Persons on hemodialysis may be at risk if chloramines are present in dialysate water. Increased methemoglobin levels and denaturation of hemoglobin have been reported in a few hemodialysis patients where dialysate water was treated with chloramines.[18] Trace chloramines are harder to remove from the dialysate water than chlorine.

Further research needs to be done to determine chronic health effects. The NAS[16] has estimated a SNARL of 0.5 mg/L for chloramines, assuming a 20% relative drinking water source contribution. Again, the drinking water source contribution could be estimated to be 90% or more, yielding a SNARL that could be up to nearly 5 times more.

DISINFECTION BY-PRODUCTS

Trihalomethanes (THMs)

Trihalomethanes present in drinking water include chloroform, bromoform, bromodichloromethane, and dibromochloromethane. These compounds are formed from the reaction of chlorine with organic matter such as humus, fulvic materials, and amides present in the water during the disinfection process.

Liver and kidney effects have been observed in rats, mice, and dogs, and decreased immune functions have been observed in mice.[19–21]

The most noted health effect reported resulting from the exposure to THMs, and in particular chloroform, is carcinogenicity. Chloroform has been found to be carcinogenic to rats and mice. The National Cancer Institute[22] reported an increased incidence of kidney tumors in male rats and liver tumors in male and female mice when chloroform was administered by gavage in a corn oil vehicle. Kidney tumors were also reported in male rats exposed to chloroform in drinking water[23] and male mice exposed to chloroform in toothpaste.[24] Liver tumors were not reported to be significantly increased in the drinking water or toothpaste studies. While chloroform has been implicated in bladder, colon, and rectal cancers in humans, the evidence is inconclusive.

The EPA has developed a cancer risk assessment for chloroform based on the incidence of liver tumors observed in mice, but the NAS[16] has suggested that the incidence of kidney tumors in rats would provide a better scientific basis for assessment. The NAS indicated that corn oil used as an indigestion medium tends to enhance the toxicity of chloroform to the liver, and therefore liver tumors may not be an appropriate end point on which to base a risk assessment for drinking water consumption.

A number of other THMs, brominated THMs, have been under testing by the National Cancer Institute, national toxicological program. Bromodichloromethane, dibromochloromethane, and bromoform have all been tested, and the results appear to be similar to what happened with chloroform and the corn oil medium. But, this does not tell us about drinking water ingestion effects.

Dibromochloromethane has also been reported to increase the incidence of liver tumors in mice.[25] Animals in this study were also given dibromochloromethane by gavage (stomach tube) in corn oil. In addition, a large number of male rats died during the course of the study.

The EPA currently has set an MCL of 0.10 mg/L for total trihalomethanes. According to the NAS report, this would correspond to an upper bound incremental lifetime cancer risk on the order of 1 in 100,000 (i.e., 10^{-5}).[16] This MCL, based primarily on treatment capabilities, was established as an Interim National Primary Drinking Water Regulation and is under reevaluation. While the carcinogenic end point of chloroform may not be at issue, the vehicle effects (corn oil versus drinking water) and tumor type are.

Chlorinated Acids, Alcohols, Aldehydes, and Ketones

The reaction of chlorine with organics in water may yield various chlorinated acids, alcohols, aldehydes, and ketones, in addition to the

THMs. EPA is evaluating the health effects of the compounds listed below for development of MCLGs:

Monochloroacetic acid	1,1-Dichloroacetone
Dichloroacetic acid	1,3-Dichloroacetone
Trichloroacetic acid	1,1,1-Trichloroacetone
Trichloroethanol	1,1,3,3-Tetrachloroacetone
Chloracetaldehyde	Pentachloroacetone
Dichloroacetaldehyde	Hexachloroacetone
Trichloroacetaldehyde (chloral)	

Currently available toxicity information on the health effects of these substances is limited. Other than an oral LD_{50} of 76 mg/kg chloroacetic acid in the rat,[26] there are no data describing the health effects of monochloroacetic acid. Reversible metabolic changes such as decreased blood glucose, lactate, pyruvate, alanine, triglyceride, and cholesterol levels have been seen in dogs receiving 150 mg/kg dichloroacetic acid for 1 to 7 days.[27]

Neurological disorders including hind limb weakness, vacuolization of white myelinated areas of the cerebrum, and various brain lesions have been observed in dogs and rats receiving doses of 50 or 125 mg/kg/day, respectively.[28]

Dichloroacetic acid has been used to treat diabetic or hyperlipoproteinemic patients. Doses of 50 mg/kg/day resulted in mild, reversible metabolic effects over a 7-day period.[29] Exposure up to 16 weeks produced a tingling sensation in the extremities and decreased strength of facial and finger muscles.[30] These effects dissipated upon cessation of treatment.

Like monochloroacetic acid, no data are available on the health effects of trichloroethanol other than an oral LD_{50} of 600 mg/kg in rats.[31] Rats receiving up to 4.5 mg/kg chloroacetaldehyde for 30 days had decreased body weight gain, hematocrit and hemoglobin counts, and increased organ-to-bodyweight ratios for brain, gonads, heart, liver, kidney, lung, and spleen.[32]

These noted effects all occurred at extremely high dose levels. There exists little information for assessing the health effects of the remainder of these by-products. Further research is needed on the effects of these compounds at the parts-per-billion levels found in drinking water.

Haloacetonitriles, Cyanogen Chloride, and Chloropicrin

Bromochloroacetonitrile (BCAN), dibromoacetonitrile (DBAN), dichloroacetonitrile (DCAN), and trichloroacetonitrile (TCAN) are also products of the reaction between chlorine and organics in water.

Observed health effects associated with exposure to these compounds include decreased organ-to-body-weight ratios in rats and fetotoxicity.

Administration of DBAN and DCAN to rats by gavage resulted in decreased body weight gain in females following 14- or 90-day exposure to 45 mg/kg/day and 65 mg/kg/day, respectively.[33] No other consistent adverse effects were noted. Doses of 55 mg/kg DBAN, DCAN, BCAN, or TCAN resulted in decreased maternal weight gain and decreased birth weight of pups exposed in utero.[34] Decreased pup survival was also reported for DCAN and TCAN. Lung adenomas have been observed in mice given 10 mg/kg TCAN or BCAN by gavage.[35]

In studies of chloropicrin, decreased body weight gain and survival have been observed in mice and rats receiving 33 or 23 mg/kg/day chloropicrin, respectively, for 108 weeks.[36] No systemic effects were noted.

Inhalation exposure in humans has resulted in pulmonary irritation and edema. A level of 4 ppm in air incapacitated soldiers in World War I. No data are available that describe the health effects of cyanogen chloride. More research is needed on the health effects of haloacetonitriles, cyanogen chloride, and chloropicrin at the parts-per-billion levels found in drinking water.

Chlorophenols

Mono-, di-, and trichlorophenol (CP, DCP, and TCP) are potential by-products of chlorination when chlorine reacts with phenolic materials. They pose common taste and odor problems in addition to having possible toxic properties.

Behavioral abnormalities and neurotoxicity have been observed in mice given 35 mg/kg 2-CP by gavage for 14 days.[37] In rats exposed to 500 mg/L 2-CP in water, no adverse effects were observed on immunological or hematological parameters after chronic exposure.[38]

Data are available to support development of an MCLG for the various chlorinated phenols. The appropriateness of immunotoxicity as an end point for developing an MCLG, however, is unclear. Other studies are needed to support the potential immunotoxicity of 2,4-dichlorophenol.

OTHER DISINFECTANTS

Other disinfectants or treatment practices have been used in drinking water disinfection. These include ozone, iodine, bromine, potassium permanganate, silver, ferrate, high pH, ionizing radiation, and ultraviolet light. The available health effects information on these substances is extremely limited.

Ozone

Ozone is used extensively in the water treatment process. It is likely that this oxidant will be used increasingly in the United States as the focus on chlorinated by-products continues. However, ozone does not leave a residual as chlorine does and therefore can pose a problem in maintenance of water quality. There may be regrowth of biological contaminants and decreased effectiveness of disinfection as the water passes through the distribution system.

Mutagenic activity has been assessed for ozone and its by-products in water. Ozone was not reported to increase mutagenic activity in a number of bacterial systems.[39] Other data on potential health effects of ozone in drinking water are not available at this time.

Iodine

Iodine as iodide has been used in treating hyperthyroidism in humans. A dose of 2–3 grains (30 mg/kg) is reported to be lethal.[1] In a 5-year study, enlarged thyroids were observed in prisoners consuming 1 to 5 mg/L iodine in drinking water for 51 months.[40]

Bromine

Bromine has been reported to induce endocrine toxicity and neurotoxicity in rats fed diets of 4 g/kg bromine.[41] Neurotoxic effects were also noted in dogs receiving 200 mg/kg bromine in their diets for six weeks.[42] In humans receiving 6 mg Br^-/kg/day for four months, no adverse effects were reported.[43]

Insufficient data are available to assess the potential health effects of the other substances. At present, no suitable data are available to determine MCLGs for any of the other disinfectants.

SUMMARY

The EPA is required to specify criteria for the disinfection of public water supplies. At present a number of issues need to be addressed, and further data on the health effects of disinfectants and their by-products are needed before the EPA can specify such criteria. Major issues to be addressed include the potential relationship of the three main disinfectants—chlorine, chlorine dioxide, and chloramine—to cardiovascular disease.

For many disinfectants and disinfection byproducts, the short-term toxicity has been well characterized. Ozone, however, is one disinfectant for which no toxicity studies are available. Studies have been performed on its by-products; ozone is unstable and does not persist in water to the consumer's tap. Longer term studies are needed for the chlorinated acids, aldehydes, alcohols and ketones, haloacetonitriles, cyanogen chloride, chloropicrin, brominated THMs, chlorine, chlorine dioxide, chlorite, and chlorate.

Other issues needing to be given attention include the consequences of the sequencing of disinfectants in the water treatment process. This needs to be studied so that water quality is highest at the lowest risk to health of by-products of disinfection. The disinfectants might react differently depending upon the sequence in which they are added during treatment, causing formation of some compounds, breakdowns of others, reformation of additional substances that react differently with the ones that already exist.

The chemistry is very complicated. The toxicity of these chemicals is also very complex. There are many studies and much money and effort being expended at this time to answer these questions.

Promulgation of regulations for these substances is extremely complicated. EPA must keep in mind not only the chemical reactions, toxicology of health effects, benefits of these compounds in controlling microorganisms and in controlling pathogenic disease outbreaks, treatment techniques, and cost feasibilities when trying to set regulatory standards—a tall order by any standards, one that could significantly effect changes in the way water is treated in the United States.

REFERENCES

1. *Drinking Water and Health, Vol. 2* (Washington, DC: National Academy of Sciences, 1980).
2. "Registry of Toxic Effects of Chemical Substances," National Institute of Occupational Safety and Health, Washington, DC, 1984.
3. Chang, J., C. Vogt, G. Sun, and A. Sun. "Effects of Acute Administration of Chlorinated Water on Liver Lipids," *Lipids* 16:336–340 (1981).
4. Cunningham, H. "Effect of Sodium Hypochlorite on the Growth of Rats and Guinea Pigs," *Am. J. Vet. Res.* 41:295–297 (1980).
5. Revis, N., P. McCauley, R. Bull, and G. Holdsworth. "Relationship of Drinking Water Contaminants to Plasma Cholesterol and Thyroid Hormone Levels in Experimental Studies," *Proc. Natl. Acad. Sci., U.S.* 83:1485–1489 (1986).
6. Lubbers, J., S. Chauan, and J. Bianchine. "Controlled Clinical Evaluations of Chlorine Dioxide, Chlorite and Chlorate in Man," *Environ. Health Pers.* 46:57–62 (1982).
7. Crump, K. "Chlorinated Drinking Water and Cancer: The Strength of the Epidemiologic Evidence," in *Water Chlorination: Environmental Impact and Health Effects, Vol. 4,* R. L. Jolley, W. A. Brungs, J. A. Cotruvo, R. B.

Cumming, J. S. Matthews, and V. A. Jacobs, Eds. (Ann Arbor, MI: Ann Arbor Science Publishers, Inc., 1983), pp. 1481–1490.
8. Stokinger, H. E. "The Halogens and the Non-Metals Boron and Silicon," in *Patty's Industrial Hygiene and Toxicology,* 3rd rev. ed., *Vol. 2B: Toxicology,* G. D. Clayton and F. E. Clayton, Eds. (New York: John Wiley & Sons, 1981), pp. 2957–2958.
9. Orme, J., D. Taylor, R. Laurie, and R. Bull. "Effects of Chlorine Dioxide on Thyroid Function in Neonatal Rats," *J. Toxicol. Environ. Health* 15:315–322 (1985).
10. Taylor, D., and R. Pfohl. "Effects of Chlorine Dioxide on Neurobehavioral Development of Rats," in *Water Chlorination: Chemistry, Environmental Impact and Health Effects, Vol. 5,* R. L. Jolley, R. J. Bull, W. P. Davis, S. Katz, M. H. Roberts, Jr., and V. A. Jacobs, Eds. (Chelsea, MI: Lewis Publishers, Inc., 1985), pp. 355–365.
11. Suh, D., M. Abdel-Rahman, and R. Bull. "Effect of Chlorine Dioxide and Its Metabolites in Drinking Water on Fetal Development in Rats," *J. Appl. Toxicol.* 3:75–79 (1985).
12. Suh, D., M. Abdel-Rahman, and R. Bull. "Biochemical Interactions of Chlorine Dioxide and Its Metabolites in Rats," *Arch. Environ. Contam. Toxicol.* 13:163–169 (1984).
13. Abdel-Rahman, M., D. Couri, and R. Bull. "Toxicity of Chlorine Dioxide in Drinking Water," *J. Am. Coll. Toxicol.* 3:277–284 (1984).
14. Moore, G., and E. Calabrese. "Toxicological Effects of Chlorite in the Mouse," *Environ. Health Pers.* 46:31–37 (1982).
15. Heffernan, W., C. Guion, and R. Bull. "Oxidative Damage to the Erythrocyte Induced by Sodium Chlorite *in vivo,*" *J. Environ. Pathol. Toxicol.* 2:1487–1499 (1979).
16. *Drinking Water and Health, Vol. 7* (Washington, DC: National Academy of Sciences, 1987).
17. Abdel-Rahman, M., D. Suh, and R. Bull. "Toxicity of Monochloramine in the Rat: An Alternative Drinking Water Disinfectant," *J. Toxicol. Environ. Health* 13:825–834 (1984).
18. Kjellstrand, C., J. Eaton, Y. Yawata, H. Swoffard, C. Kolpin, T. Buselmeier, B. von Hartitzsch, and H. Jacob. "Hemolysis in Dialysized Patients Caused by Chloramines," *Nephron* 13:427–433 (1974).
19. Jorgenson, T., and C. Rushbrook. "Effects of Chloroform in the Drinking Water of Rats and Mice: Ninety-Day Subacute Toxicity Study," EPA-600/180–030 (1980).
20. Heywood, R., R. Sortwell, P. Noel, A. Street, D. Prentice, F. Roe, P. Wadsworth, A. Warden, and N. Yan Abbe. "Safety Evaluation of Toothpaste Containing Chloroform. III. Long-Term Study in Beagle Dogs," *J. Environ. Pathol. Toxicol.* 2:835–851 (1979).
21. Munson, A., L. Sain, V. Sanders, B. Kauffman, W. White, D. Page, D. Barnes, and J. Borzelleca. "Toxicology of Organic Drinking Water Contaminants: Trichloromethane, Bromodichloromethane, Dibromochloromethane and Tribromomethane," *Environ. Health Pers.* 46:117–126 (1981).

22. "Report on the Carcinogenesis Bioassay of Chloroform," National Cancer Institute, Washington, DC (1976).
23. Jorgenson, R., E. Meierhenry, C. Rushbrook, R. Bull, and M. Robinson. "Carcinogenicity of Chloroform in Drinking Water to Male Osborne-Mendel Rats and Female $B6C3F_1$ Mice," *Fund. Appl. Toxicol.* 5:760-769 (1985).
24. Roe, F., A. Palmer, A. Warden, and N. Van Abbe. "Safety Evaluation of Toothpaste Containing Chloroform. I. Long-Term Studies in Mice," *J. Environ. Pathol. Toxicol.* 2:799–819 (1979).
25. "Toxicology and Carcinogenesis Studies of Chlorodibromomethane in F344 Rats and $B6C3F_1$ Mice (Gavage Studies)," RS No. 282, (Washington, DC: National Toxicology Program, 1985).
26. Woodward, G., S. Lange, W. Nelson, and H. Colvery. "The Acute Oral Toxicity of Acetic, Chloracetic, Dichloroacetic and Trichloroacetic Acids," *J. Ind. Hyg. Toxicol.* 23:78–82 (1941).
27. Ribes, G., G. Valette, and M. Loubathieres-Marian. "Metabolic Effects of Sodium Dichloroacetate in Normal and Diabetic Dogs," *Diabetes* 28:852–857 (1979).
28. Katz, R., C. Tai, R. Diener, R. McConnell, and D. Semanick. "Dichloroacetate Sodium: 3-Month Oral Toxicity Studies in Rats and Dogs," *Toxicol. Appl. Pharmacol.* 57:273–287 (1986).
29. Stacpoole, P., G. Moore, and D. Kornauser. "Metabolic Effects of Dichloroacetate in Patients with Diabetes Mellitus and Hyperlipid Proteinemia," *New England J. Med.* 298:526–530 (1978).
30. Stacpoole, P., G. Moore, and D. Kornauser. "Toxicity of Chronic Dichloroacetate," *New England J. Med.* 300:372 (1979).
31. "The Toxic Substances List," U.S. Department of Health, Education and Welfare, Washington, DC, 1974.
32. Lawrence, W., E. Dillingham, J. Turner, and J. Antion. "Toxicity Profile of Chloroacetaldehyde," *J. Pharm. Sci.* 61:19–25 (1972).
33. Hayes, J., L. Condie, and J. Borzelleca. "Toxicology of Haloacetonitriles," *Environ. Health Pers.* 69:183–202 (1986).
34. Smith, M., H. Zenick, and E. George. "Reproductive Toxicology of Disinfection By-Products," *Environ. Health Pers.* 69:177–182 (1986).
35. Bull, R., and M. Robinson. "Carcinogenic Activity of Haloacetonitrile and Haloacetone Derivatives in Mouse Skin and Lung," in *Water Chlorination: Chemistry, Environmental Impact and Health Effects, Vol. 5,* R. L. Jolley, R. J. Bull, W. P. Davis, S. Katz, M. H. Roberts, Jr., and V. A. Jacobs, Eds. (Chelsea, MI: Lewis Publishers, Inc., 1985).
36. "Carcinogenesis Bioassay of Chloropicrin," RS No. 65, (Washington, DC: National Cancer Institute, 1978).
37. Kallman, M., E. Coleman, and J. Borzelleca. "Behavioral Toxicity of 2-Chlorophenol in Adult Mice," *Abstr. Fed. Proc.* 41:1580 (1982).
38. Exon, J., and L. Koller. "Effects of Chlorinated Phenols on Immunity in Rats," *Int. J. Immunopharmacol.* 5:131–134 (1983).
39. Cotruvo, J., V. Simmon, and R. Spanggard. "Investigation of Mutagenic

Effects of Products of Ozonation Reactions in Water," *Ann. N.Y. Acad. Sci.* 298:124–140 (1977).
40. Thomas, W., A. Black, G. Freund, and R. Hinman. "Iodine Disinfection of Water," *Arch. Environ. Health* 19:124–128 (1969).
41. Van Logden, M., M. Nolthuis, A. Rauhs, R. Kroes, E. Den Tonkelaar, H. Berkvens, and G. Van Esch. "Semichronic Toxicity Study in Sodium Bromide in Rats," *Toxicol.* 2:257–267 (1974).
42. Rosenblum, I. "Bromide Intoxication: Production of Experimental Interaction in Dogs," *J. Pharmacol. Exp. Ther.* 122:379–384 (1958).
43. Flinn, F. "The Appearance of Blood Bromide After Oral Ingestion," *J. Lab. Clin. Med.* 26:1325–1329 (1941).

CHAPTER 8

National Primary Drinking Water Regulations for Additional Contaminants to Be Regulated by 1989

Edward V. Ohanian

INTRODUCTION

The Safe Drinking Water Act (SDWA) of 1974 requires the U.S. Environmental Protection Agency (EPA) to publish National Primary Drinking Water Regulations (NPDWRs) for contaminants that may pose adverse human health effects. On June 19, 1986, President Reagan signed the 1986 Amendments to the SDWA. The amendments state that:

1. EPA must propose and promulgate maximum contaminant level goals (MCLGs) (unenforceable health goals) and maximum contaminant levels (MCLs) (enforceable standards) simultaneously
2. EPA must regulate 9 contaminants within one year of enactment, another 40 within two years, and 34 additional contaminants within three years (for a total of 83)
3. EPA has the option of substituting up to seven other contaminants for those listed by Congress if it finds this will provide greater health protection
4. in addition to the 83 contaminants, at least 25 more primary drinking water standards will be required by 1991, with 25 more standards expected every three years thereafter

5. EPA must prepare a report to Congress on comparative health risks of raw water contamination versus contamination by treatment chemicals (e.g., disinfection by-products) within 18 months of enactment

In order to execute EPA's responsibilities under the amendments, Congress authorized $170 million for fiscal year 1987.

Section 1412(b)(3)(A) of the SDWA Amendments requires the administrator of the EPA to publish MCLGs and promulgate NPDWRs for each contaminant, which in the judgment of the administrator may have adverse effects on public health and is known or anticipated to occur in public water systems. The MCLG is nonenforceable and is set at a level at which no known or anticipated adverse health effects in humans occur and which allows for an adequate margin of safety. Factors considered in setting an MCLG include health effects data and sources of exposure other than drinking water.

Drinking Water Criteria Documents (CDs) have been or are being prepared for each contaminant to be regulated. These provide the health effects basis to be considered in establishing the MCLG. To achieve this objective, data on pharmacokinetics, human exposure, acute and chronic toxicity to animals and humans, epidemiology, and mechanisms of toxicity are evaluated. Specific emphasis is placed on literature data providing dose-response information. Thus, while the literature search and evaluation performed in support of each CD is comprehensive, only the reports considered most pertinent in the derivation of the MCLG are cited in the documents.

CONTAMINANTS SCHEDULED FOR REGULATION BY 1989

As discussed above, the 1986 Amendments to the SDWA established an agenda for regulation of 83 contaminants over a three-year period. The contaminants currently scheduled for regulation by 1987, 1988, and 1989 are summarized in Table 1. Proposed and/or final rulemaking has already been published in the *Federal Register* for those contaminants listed for regulation by 1987 and 1988. Thus, the next group anticipated to enter the regulatory process are the 34 contaminants to be regulated by 1989. Included are representatives from all five categories of contaminants, including the first NPDWRs for radionuclides. It should be noted, however, that the EPA may make up to seven substitutions to the list in Table 1 if it is determined that regulation of the substitute is likely to be more protective of human health.

Table 1. Contaminants Scheduled for Regulation by 1989 Under the 1986 Amendments to the Safe Drinking Water Act

	By 1987 (9 contaminants)	
VOLATILE ORGANIC CHEMICALS		
Benzene	1,2-Dichloroethane	Trichloroethylene
Carbon tetrachloride	1,1-Dichloroethylene	Vinyl chloride
p-Dichlorobenzene	1,1,1-Trichloroethane	
INORGANICS		
Fluoride		
	By 1988 (40 contaminants)	
VOLATILE ORGANIC CHEMICALS		
Chlorobenzene	*cis*-1,2-Dichloroethylene	*trans*-1,2-Dichloroethylene
o-Dichlorobenzene		
MICROBIALS AND TURBIDITY		
Giardia	Turbidity	Viruses
Total coliforms		
INORGANICS		
Arsenic	Chromium	Nitrate
Asbestos	Copper	Selenium
Barium	Lead	
Cadmium	Mercury	
ORGANICS		
Acrylamide	1,2-Dichloropropane	Methoxychlor
Alachlor	Epichlorohydrin	PCBs
Aldicarb	Ethylbenzene	Pentachlorophenol
Carbofuran	Ethylene dibromide (EDB)	Styrene
Chlordane		Toluene
2-4-D	Heptachlor	Toxaphene
Dibromochloropropane (DBCP)	Heptachlor epoxide	2,4,5-TP
	Lindane	Xylene
	By 1989 (34 contaminants)	
VOLATILE ORGANIC CHEMICALS		
Methylene chloride (Dichloromethane)	Trichlorobenzene	
MICROBIALS AND TURBIDITY		
Legionella	Standard plate count	
INORGANICS		
Antimony	Cyanide	Sulfate
Beryllium	Nickel	Thallium
ORGANICS		
Acrylonitrile	Endothall	Phthalates
Adipates	Endrin	Picloram
Atrazine	Glyphosate	Simazine
Dalapon	Hexachlorocyclopentadiene	2,3,7,8-TCDD (Dioxin)
Dinoseb		1,1,2-Trichloroethane
Diquat	PAHs	Vydate (Oxamyl)
RADIONUCLIDES		
Radium-226	Uranium	Photon radioactivity
Radium-228	Gross α particle activity	
Radon	β particle activity	

SCIENTIFIC BASIS FOR DEVELOPING REGULATIONS

Criteria Document Development Process

In order to establish MCLGs and MCLs, EPA must first assemble a database for evaluating potential human health effects as they relate to the presence of each contaminant in drinking water. This database, as developed by the Office of Drinking Water (ODW), is assembled in the CD. The development of a CD follows a stepwise sequence that parallels the regulatory process, both of which are estimated to require approximately 36 months from the time of chemical identification. Both the regulatory and CD development processes are summarized in Figure 1.

At the time of chemical identification, an Advance Notice of Proposed Rulemaking (ANPRM) is published in the *Federal Register,* thereby beginning the regulatory process. The ANPRM allows for public comments, meetings, and workshops before development and issuance of the proposed MCLG and MCL. During the first year of the process, the CD rough draft and rough external review draft are prepared. These drafts are systematically presented to experts within the EPA for review and comment on the adequacy of the database and risk assessments proposed as the health basis for the MCLG. Before preparing the final draft CD, the external review draft is submitted to a panel of recognized experts outside of EPA for a thorough, critical review. Following this revision step, the final draft CD serves as the technical support document providing the health basis for the proposed MCLG published in the *Federal Register.* Following the review and comment period on both the final draft CD and the proposed MCLG, all inputs are considered in producing the final version of the CD. The final CD serves as the technical support document providing the health basis for the final rule promulgating the MCLG.

During the entire regulatory process, other documents are prepared by ODW dealing with analytical methods, treatment technologies, human exposure potential and cost-benefit analyses for each contaminant. These other documents are used in deriving the proposed and final MCL, which is to be set as close to the MCLG (i.e., health goal) as is feasible. Thus, by providing the health basis for deriving the MCLG, the CD serves as the keystone to the regulatory process for drinking water contaminants. The MCL is then derived from the MCLG by evaluating the feasibility of regulation at the health goal level when the availability and cost of analytical methods and treatment technologies are considered. If at any time during the CD and regulatory processes data become available which indicate that the proposed MCLG and/or MCL are inadequate to protect human safety, these regulations may be amended. Addi-

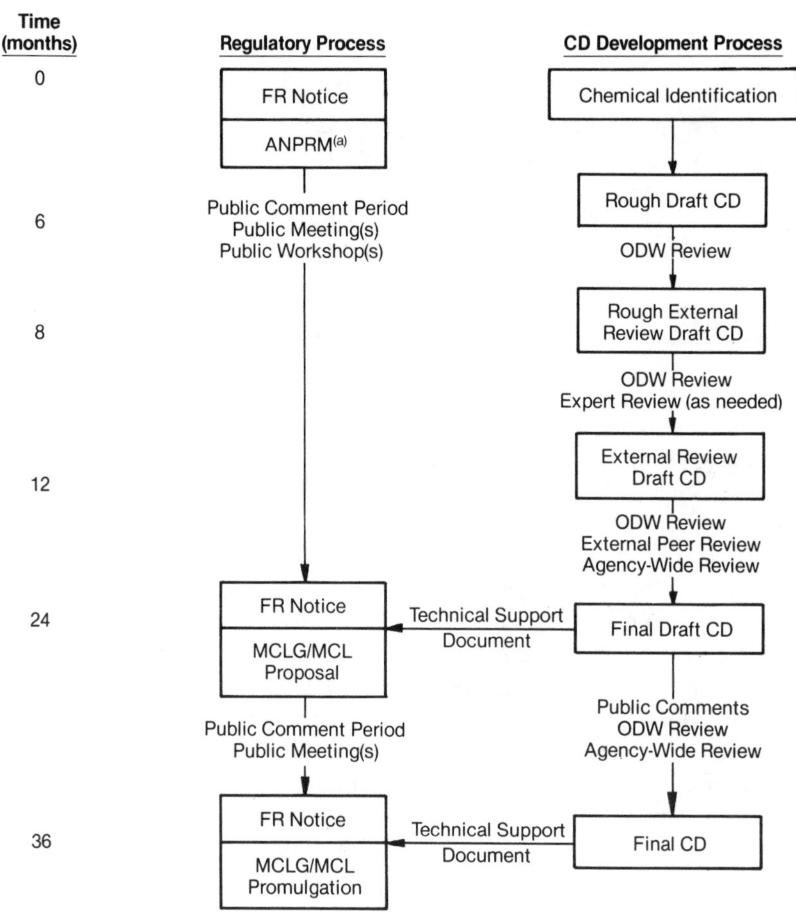

(a) Advanced Notice of Proposed Rulemaking

Figure 1. The ODW Criteria Document development process.

tionally, the relevance and adequacy of the NPDWRs are to be reviewed at least every three years and revised regulations promulgated when appropriate.

Major Elements of a Criteria Document

By definition, the EPA is required to promulgate NPDWRs for contaminants that are known or anticipated to occur in drinking water and that may cause an adverse human health effect. The primary objectives of CDs are, therefore, to establish core information based on health effects of chemicals in drinking water and to compile and evaluate data for providing the qualitative and quantitative health effects basis for MCLGs. Each CD consists of nine chapters: 1) summary, 2) physical and chemical properties, 3) toxicokinetics, 4) human exposure, 5) health effects in animals, 6) health effects in humans, 7) mechanism of toxicity, 8) quantification of toxicological effects, and 9) references.

Quantification of Toxicological Effects

Of the nine chapters of the CD, it is the Quantification of Toxicological Effects (QTE) section that integrates the key health effects information and provides the basis for the proposed MCLG. The objective of risk assessments performed in this section is to define the level at which no known or anticipated adverse effects on human health may occur, while allowing for an adequate margin of safety to protect more sensitive individuals. The QTE for a chemical consists of separate assessments of noncarcinogenic and carcinogenic health effects. Chemicals that do not produce carcinogenic effects are believed to have a threshold dose below which no adverse, noncarcinogenic health effects occur. Carcinogens are assumed to act without a threshold (i.e., there is no exposure level that is assumed to be without some level of health risk). For nonchemical drinking water contaminants (e.g., microorganisms, turbidity, etc.), the organizational structure and risk assessment procedures are modified, as required by the contaminant's health effects properties. These case-by-case exceptions to the QTE's structure, however, will not be addressed in this chapter.

Noncarcinogenic Effects

In the quantification of noncarcinogenic effects, a reference dose (RfD)—formerly termed the acceptable daily intake (ADI)—is calculated. The RfD is an estimate (with an uncertainty spanning perhaps an order of magnitude) of the daily exposure to the human population (including sensitive subgroups) that is likely to be without an appreciable risk of deleterious health effects during a lifetime. The RfD is derived from a no-observed-adverse-effect level (NOAEL) or a lowest-observed-adverse-effect level (LOAEL), identified from a subchronic or chronic study, and divided by an uncertainty factor(s). The RfD is calculated as follows:

$$\text{RfD} = \frac{\text{(NOAEL or LOAEL)}}{\text{Uncertainty Factor(s)}} = \text{mg/kg bw/day}$$

The use of an uncertainty factor(s) in the calculation accounts for possible intra- and interspecies variability that may exist when extrapolating exposure levels from specific animal or human studies to the general population. Uncertainty factor selection is based upon professional judgment, while considering the entire database of toxicological effects for the chemical. To ensure that uncertainty factors are selected and applied in a consistent manner, ODW employs a modification to the guidelines proposed by the National Academy of Sciences (NAS). These guidelines are as follows:

- An uncertainty factor of 10 is generally used when good chronic or subchronic human exposure data identifying a NOAEL are available and are supported by good chronic or subchronic toxicity data in other species.
- An uncertainty factor of 100 is generally used when good chronic toxicity data identifying a NOAEL are available for one or more animal species (and human data are not available), or when good chronic or subchronic toxicity data identifying a LOAEL in humans are available.
- An uncertainty factor of 1000 is generally used when limited or incomplete chronic or subchronic toxicity data are available, or when good chronic or subchronic toxicity data identifying a LOAEL (but not a NOAEL) for one or more animal species are available.

Additional considerations, not incorporated in the NAS/ODW guidelines, may necessitate the use of an additional uncertainty factor of 1 to 10. These other considerations include: the use of a less than lifetime study for deriving the RfD, the significance of the adverse health effect, pharmacokinetic factors, and the counterbalancing of beneficial effects (e.g., essential nutrients).

From the RfD, a drinking water equivalent level (DWEL) is calculated. The DWEL represents a medium-specific (e.g., drinking water) lifetime exposure at which adverse, noncarcinogenic health effects are not anticipated to occur. The DWEL assumes 100% of the human exposure is derived from drinking water and provides the noncarcinogenic health effects basis for establishing the MCLG. For ingestion data, the DWEL is derived as follows:

$$\text{DWEL} = \frac{\text{(RfD)} \times \text{(Body Weight in kg)}}{\text{Drinking Water Volume in L/day}} = \text{mg/L}$$

where Body weight = assumed to be 70 kg for an adult
Drinking water volume = assumed to be 2 liters per day for an adult

Carcinogenic Effects

If toxicological evidence leads to the classification of the contaminant as a known or probable human carcinogen, mathematical models are used to calculate drinking water concentrations associated with estimated excess cancer risk levels (e.g., 10^{-4}, 10^{-5}, 10^{-6}). A risk of 10^{-4}, for example, indicates a probability of one additional case of cancer for every 10,000 people exposed; a risk of 10^{-5} indicates one additional cancer per 100,000 exposed individuals. The data used in these risk estimates usually come from lifetime exposure studies in animals. To predict the risk for humans, these animal doses must be converted to equivalent human doses. This conversion includes correction for noncontinuous animal exposure, less-than lifetime studies and differences in animal/human size. The factor that compensates for the size difference is the cube root of the ratio of the animal and human body weights. It is assumed that the average adult human body weight is 70 kg and that the average water consumption of an adult human is 2 liters of water per day.

For contaminants with a carcinogenic potential, drinking water concentrations are correlated with the carcinogenic risk estimates by employing a cancer potency (unit risk) value together with the assumption for lifetime exposure via ingestion of water. The cancer unit risk is usually derived from a linearized multistage model with a 95% upper confidence limit providing a low-dose estimate. The true cancer risk to humans, while not experimentally measurable, is not likely to exceed this upper-limit estimate and, in fact, may be lower. Excess cancer risk estimates may also be calculated using the one-hit, Weibull, logit, and probit models. There is a limited basis in the current understanding of the biological mechanisms involved in cancer to suggest that any one of these models predicts risk more accurately than another. Because each model is based upon differing assumptions, the estimates derived for each model can differ by several orders of magnitude.

The scientific database used to calculate and support cancer risk rate estimates has an inherent uncertainty due to systematic and random errors in scientific measurement. In most cases, only cancer studies using experimental animals have been performed. Thus, there is uncertainty when data are extrapolated to humans. When developing cancer risk rate estimates, several other areas of uncertainty exist, such as the incomplete knowledge concerning the health effects of contaminants in drinking water, the impact of the experimental animal's age, sex and species, the nature of target organ system(s) examined and the actual rate of exposure to the internal targets in experimental animals or humans. Dose-response data are usually available only for high levels of exposure, not for the lower levels of exposure closer to where a standard may be set. When there is exposure to more than one contaminant, additional uncertainty results from a lack of information about possible synergistic or antagonistic effects. All of these uncertainties support use of the more conservative approach (e.g., linearized multistage model) for estimating cancer risk rates.

Development of MCLGs

From the assembled information on noncarcinogenic and carcinogenic effects, ODW employs a three-category approach for development of the MCLG for each contaminant. This approach is summarized in Table 2. The primary consideration is whether the contaminant has been classified as a carcinogen. The EPA classification scheme for carcinogens is provided in Table 3. Once the carcinogenic classification has been established, the MCLG derivation is relatively straightforward. From Table 2, the MCLG for a "strong" carcinogen is set at zero because a carcinogen is assumed to act without a threshold. The MCLG for an "equivocal" carcinogen is calculated based upon either the noncarcinogenic (DWEL approach) or carcinogenic (cancer risk approach) data based upon the availability of information and professional judgments made on a case-by-case basis. When evidence of carcinogenicity is "inadequate or lacking," the MCLG is derived from the noncarcinogenic data. Here, the DWEL and the percentage of the estimated total human exposure deriving from drinking water (i.e., relative source contribution) are multiplied

Table 2. Three-Category Approach for Developing MCLGs

Evidence of Carcinogenicity	Classification	MCLG
Strong	EPA Group A or B[a]	MCLG is set at zero
Equivocal	EPA Group C	MCLG is derived by: (a) DWEL approach with additional safety factor or (b) Set within the 10^{-5} to 10^{-6} cancer risk range
Inadequate or negative	EPA Group D or E	MCLG is derived by standard DWEL approach; MCLG = (DWEL) (% of drinking water contribution)

Note: DWEL = drinking water equivalent level; EPA = Environmental Protection Agency; IARC = International Agency for Research on Cancer; MCLG = maximum contaminant level goal.

[a] Consult Table 3 for detail.

Table 3. EPA Classification of Carcinogens

Group	Evidence of Carcinogenicity
A	Human carcinogen (sufficient evidence from epidemiological studies)
B	Probable human carcinogen
B_1	At least limited evidence of carcinogenicity to humans
B_2	Usually a combination of sufficient evidence in animals and inadequate data in humans
C	Possible human carcinogen (limited evidence of carcinogenicity in animals in the absence of human data)
D	Not classified (inadequate animal evidence of carcinogenicity)
E	No evidence of carcinogenicity for humans (no evidence for carcinogenicity in at least two adequate animal species or in both epidemiological and animal studies)

Note: EPA = Environmental Protection Agency.

to derive the MCLG. This value then defines the concentration of the contaminant that can remain in drinking water and result in no anticipated adverse health effects over a person's lifetime, even with other estimated sources of exposure such as food, air, and so on.

RESEARCH NEEDS FOR DEVELOPING REGULATIONS

While ODW is engaged in the most detailed and comprehensive assessment of drinking water quality specifications ever attempted and considering that progress has been made to fulfill a substantial portion of this mandate, it has become evident that quantitative dose-response information on the noncarcinogenic and carcinogenic effects for several of the contaminants to be regulated are not available. This leaves the drinking water CDs incomplete and thwarts attempts to derive meaningful drinking water standards. Of particular concern is the evaluation of the carcinogenic potential of contaminants. Provisions of the SDWA require that a finding of carcinogenicity (no safe threshold) for a particular contaminant leads to the conclusion that the MCLG must be set a zero. Obviously, such findings have widespread impact and must be validated to withstand the scientific and legal challenges that will be encountered during the regulatory review process.

Although ODW's priority research needs are for good quality dose-response toxicity data (e.g., data from long-term, lifetime, animal, or human studies), there are additional research needs that warrant emphasis. Even when there are good dose-response data for a contaminant, replicate data from the same and other species are highly desirable. Uncertainties associated with intra- and interspecies variation often force

ODW to adopt conservative risk assessment policies for deriving the MCLG values. Obviously, the more comprehensive the database that exists for a chemical, the greater the certainty in making risk extrapolations to humans. While the ODW strives to estimate realistic, safe human exposure levels, they must retain the "margin of safety" principle incorporated in the SDWA. This leads to criticisms and controversy that could be reduced by more comprehensive toxicity databases.

Major sources of uncertainty worthy of emphasis in the risk assessment process are concerns over the most sensitive toxicity endpoint and most sensitive subpopulation(s). A drinking water standard should provide public health protection from all adverse effects. With a heavy reliance on toxicity data from the open literature, there is often uncertainty as to whether the most sensitive toxicity endpoints have been evaluated. Many toxicity researchers focus their efforts within their academic disciplines (e.g., neurotoxicity, development effects, renal effects, etc.). This often results in toxicity data for a variety of chemicals for that specific toxicity endpoint. Information gaps on other potentially relevant toxicity endpoints often exist. Thus, there are a number of information voids for drinking water contaminants that if filled, even with negative data, would provide added confidence during risk assessments. Similarly, comparative toxicity or pharmacokinetic studies that identify effects associated with different species, age, sex, and so on are valuable in selecting the most appropriate human model and identifying the most sensitive subpopulations that require protection.

The ODW also has a keen interest in research on a variety of scientific issues associated with performing risk assessments. Briefly, these include:

1. Improving the use of pharmacokinetic information in evaluating dose-response data.
2. Improving procedures for estimating human inhalation exposures due to contaminants in drinking water (e.g., relevant to showering, cooking, etc.).
3. Developing a consensus for risk assessment procedures for estimating human risk levels via dermal exposures (e.g., washing, bathing, etc.) from drinking water contaminants. As a subset of this need is the availability of dermal dose-response data for cost contaminants. This information is generally lacking, especially for exposures and toxic endpoints outside the occupational setting. Whether effects from dermal exposures are significant relative to other exposures, is often debatable, but remains largely unknown for many contaminants.

Finally, it is evident that ODW's research needs are extensive and transcend the needs of other EPA offices (Air, Hazardous Waste, Toxic

Substances, etc.), as well as other regulatory and health agencies at the federal, state, and local level. While EPA has some limited research funding capabilities, much of ODW's toxicity data needs must be met by research supported by other private or public sector sponsors. Where modest adjustments can be made in toxicity study protocols to meet multiple needs, innovative and cooperative funding arrangements should be sought to support research. Regardless of how individual research projects are funded, ODW requests that consideration always be given to no-cost or low-cost adjustments to testing protocols that extend the utility of the results to better meet ODW's needs.

Thus, the primary focus of future experimental, clinical, and epidemiological research undertaken for a regulatory agency such as EPA must be able to provide the essential quantitative dose-response data needed to support policymaking and regulatory decisions required under the SDWA Amendments. Research efforts should be directed toward the identification of the significant toxic endpoints and the quantification of the dose-response relationships to be used in the generation of MCLGs.

CHAPTER 9

Chemicals for Regulation by 1991 and Beyond

Joseph A. Cotruvo and Marlene Regelski

Phase VI will begin the first cycle of the requirement of the Safe Drinking Water Act of 1974 (SDWA) to regulate or reassess 25 additional substances every three years. The target date for these first 25 additional chemicals for final promulgation is January 1991. The regulatory effort will include development of drinking water priority lists (DWPLs) to be published triennially.

Substances for future regulatory consideration will include those chemicals listed pursuant to the SDWA Section 1428 (wellhead protection), other CERCLA (Comprehensive Environmental Response, Compensation, and Liability Act) Section 101 (14) substances, and substances not included on the first drinking water priority list because of data limitations.

Publication of the first DWPL occurred in January of 1988. MCLs are required to be set for at least 25 substances within 36 months of that publication. This process is scheduled to be repeated every three years.

These lists will be derived from a number of organizations with which we communicate to identify the best list of candidate substances. These organizations include, within the U.S. Environmental Protection Agency (EPA), for example, those offices in Superfund and Hazardous Waste, Groundwater, Water Quality, Pesticides, and Toxic Substances. In addition, EPA will consult with outside agency groups such as the National Toxicology Program.

Since this Phase VI list will include substances for which we have not had sufficient data, we necessarily must take into consideration the availability of health effects data, analytic methodology, occurrence data, treatment technology, and how much data will be sufficient to determine regulatory levels.

We will need to look across the board at private wells, additive substances, surface waters, waste waters, environmental chemistry of substances, the mobility of these substances in the environment, and mechanisms for their getting into drinking water, either in the ground or on the surface. Patterns of use of these compounds and their production, properties of biodegradation and absorption, and the amounts in which they are found will need to be studied.

Subsequent lists and additional substances to be regulated have a great potential for being priorities across EPA programs throughout the agency. This will afford the opportunity to consolidate the decision processes of all those programs.

CHAPTER 10

Office of Drinking Water's Health Advisory Program

Edward V. Ohanian

INTRODUCTION

The Office of Drinking Water's (ODW) Health Advisory (HA) Program provides information on health effects, analytical methodology and treatment technology that would be useful in dealing with the contamination of drinking water. Health Advisory documents include nonregulatory concentrations of contaminants in drinking water at which adverse effects would not be anticipated to occur over specific durations of exposure. A margin of safety is included in the HA values to protect sensitive members of the population. The HA values are not legally enforceable federal standards. They are subject to change as new and better information becomes available. The HAs are offered as informal technical guidance to assist federal, state and local officials responsible for the protection of public health.

The HA values are developed from data describing noncarcinogenic endpoints of toxicity. They do not quantitatively incorporate any potential carcinogenic risk from such exposure. For those chemicals which are known or probable human carcinogens according to the U.S. Environmental Protection Agency (EPA) classification scheme (Group A or B), One-Day, Ten-Day and Longer-Term HAs may be derived, with attendant caveats. Health Advisories for lifetime exposures are not recommended for carcinogens. Rather, projected excess lifetime cancer risks (unit risk) are provided to give an estimate of the concentrations of the contaminant that may pose a carcinogenic risk to humans. These hypothetical estimates are usually presented as upper 95% confidence limits derived from the linearized multistage model, which is considered to be unlikely to underestimate the probable true cancer risk. Excess cancer risk estimates may also be calculated using the one-hit, Weibull, logit, and probit models. Since these models are based on different assumptions, the resulting risk estimates may differ by several orders of magnitude.

When an ODW Health Effects Criteria Document is available, the HA is based upon information presented in the Criteria Document. Individuals desiring further information on the toxicological database or rationale for risk characterization should consult the Criteria Document. Criteria Documents and HAs are available for review at each EPA Regional Office or Drinking Water counterpart (e.g., Water Supply Branch or Drinking Water Branch) or, for a fee, from the National Technical Information Service, U.S. Department of Commerce, 5285 Port Royal Road, Springfield, Virginia.

ELEMENTS OF THE HEALTH ADVISORY PROGRAM

Functions of the Office of Drinking Water

The HA Program, as conducted by the ODW, is an on-going, multiphased program designed to provide the most currently available information on potential or known drinking water contaminants in a timely manner. Earlier phases of the program, which began in 1979, have resulted in the preparation of HAs for approximately 50 drinking water contaminants. Currently HAs for approximately 60 National Pesticide Survey (NPS) analytes and 50 unregulated volatile organic chemicals (VOCs) are in the process of preparation. The Department of the Army has also entered into a cooperative program with the EPA to develop HAs for various munition chemicals having the potential to contaminate drinking waters.

After each HA has been prepared and becomes available for public use, it is entered into a computer-based HA Registry, and executive summaries are entered into the Integrated Risk Information System (IRIS), a database for information on risk assessment and risk management throughout EPA.

In addition to preparing and dispersing the information contained in the HAs, the ODW is continually involved in updating the existing HAs, preparing and streamlining procedures to assure timely and accurate response to emergency and noncritical requests for information, and initiating information-sharing and toxicological support programs between the states and the EPA.

Finally, the ODW has undertaken a program to conduct three-day workshops to provide users of the HAs and other drinking water-related documents with information on the philosophy, methodology, and application of risk assessment and risk management decisionmaking as it exists at all levels of government.

Health Advisory Development Process

The development of an ODW HA document is a step-by-step process estimated to require approximately 12 months from the time of identification of the chemical to issuance of a final HA. This process is illustrated in Figure 1. The development process includes a minimum of four separate review steps by individuals within and outside of the EPA to ensure the quality and accuracy of the final HA.

The initial review process takes place within the ODW itself, utilizing a toxicological review panel to provide a critical review of the rough draft's content, assumptions, and conclusions. The rough external review draft results from this critique. This is submitted to a number of individuals within the EPA who are experts on either the chemical in question or the various subdisciplines contained in the draft HA (i.e., pharmacokinetics, toxicology, carcinogenicity, etc.). During each step in the review process, the draft may be resubmitted to a reviewer, as needed, to reassess any changes deemed necessary.

The draft HA is next submitted to an external peer review panel of recognized experts as well as to an agency-wide review before the final draft HA is prepared. This final draft is then presented for public comment. A final HA is not issued for public use until all phases of the review process have been satisfied.

HEALTH ADVISORIES

Legal Status of Health Advisories

Health Advisories are neither legally enforceable standards nor are they issued as official regulations. They may or may not lead to the issuance of national standards or maximum contaminant levels (MCLs). The HAs do not condone the presence of contaminants in drinking water; rather, they are prepared to provide specific advice on the levels of contaminants as they relate to possible health effects. They describe

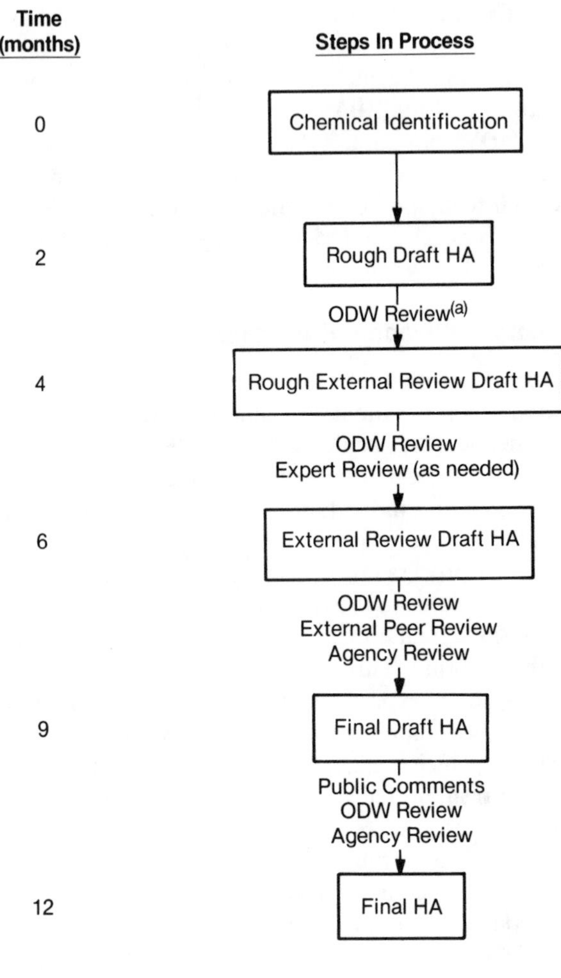

(a) Toxicological Review Panel

Figure 1. Office of Drinking Water's Health Advisory development process.

concentrations of contaminants in drinking water at which adverse non-carcinogenic effects would not be anticipated to occur following one-day, ten-day, longer-term, or lifetime exposures. The HAs are subject to change as new and better information becomes available.

Content of Health Advisories

Health Advisories are designed to present, in a capsular or "bullet" format, essential background information on the chemical. This is done to provide the user with a concise but complete profile of the chemical. For the purpose of consistency and user convenience, the HAs are prepared in the standardized format outlined in Table 1.

Following a brief standard introduction explaining the HA Program, the general information and properties of the specific chemical are presented. To assist in chemical identification, the physical and chemical properties are given in table format along with the various synonyms and uses. Sources of exposure and environmental fate are included to assist in determining the extent of possible human exposure. All available information on the pharmacokinetic properties of the chemical are added with particular emphasis on the chemical's absorptive properties and known metabolites. When data is available on health effects in humans, all

Table 1. ODW Health Advisory Content

I. General Introduction

II. General Information and Properties
- Synonyms
- Uses
- Properties
- Sources of Exposure
- Environmental Fate

III. Pharmacokinetics
- Absorption
- Distribution
- Biotransformation
- Excretion

IV. Health Effects
- Humans
- Animals
  - –Short-Term Exposure
  - –Longer-Term Exposure
- Developmental/Reproductive/Mutagenic/Carcinogenic Effects

V. Quantification of Toxicological Effects
- One-Day Health Advisory
- Ten-Day Health Advisory
- Longer-Term Health Advisory
- Lifetime Health Advisory
- Evaluation of Carcinogenic Potential

VI. Other Criteria, Guidances and Standards

VII. References

Note: ODW = Office of Drinking Water.

pertinent details are reported, to include dose and mode of exposure and the effects resulting from acute and chronic exposures.

As the bulk of information on the toxic effects of chemicals is usually derived from animal studies, this section of an HA is often the most comprehensive. If available, the animal data will include both short-term and longer-term exposure studies, reproductive and developmental effects, mutagenicity and carcinogenicity data as well as any available dermal and ocular effects information. However, since an HA is intended to be a brief guidance document, only those studies deemed most pertinent to its presence as a contaminant in drinking water are included. Thus, the HA is not meant to be a review of all available data on the specific chemical.

The next section of the HA, Quantification of Toxicological Effects, presents the rationale used in selecting the studies for development of the One-Day, Ten-Day, Longer-Term, and Lifetime HA values. These values are developed from data describing only noncarcinogenic endpoints of toxicity. If a chemical is determined to be a known or probable human carcinogen, the HA document will include carcinogenic potency factors and drinking water concentrations estimated to represent excess lifetime cancer risks.

Other known criteria, guidance or published standards are also included in the HA document as an additional means of evaluating the status of the contamination. Finally, analytical methodology and treatment technologies are included to assist the user in assessing the situation and making the appropriate public health management decisions. Should the user require additional information, a list of the cited references is included. If the HA is based upon an existing Criteria Document, it is also referenced. Additionally, the user may contact an EPA Regional Office as well as the ODW for further assistance.

Preferred Data for Health Advisory Development

In deriving the HA values, specific types of data are considered most pertinent to each phase of the process. These data may be subdivided into three categories as indicated in Table 2. The following explains how these data are selected to derive each of the HA values.

Duration of Exposure

One-Day Health Advisory. The One-Day HA is calculated for a 10-kg child and assumes a single acute exposure to the chemical. It is generally derived from a study of less than seven days' duration.

Table 2. Preferred Data for HA Development

Duration of exposure
One-Day HA: 1 to 5 (successive) daily doses
Ten-Day HA: 7 to 30 (successive) daily doses
Longer-Term HA: Subchronic (90-day) to one-year study
Lifetime HA:
Chronic study
Subchronic study (with added uncertainty factor)
Route of Administration
Oral: Drinking water, gavage, or diet
Inhalation
Subcutaneous or intraperitoneal
Test species
Human
Appropriate animal model
Most sensitive species

Note: HA = Health Advisory.

Ten-Day Health Advisory. The Ten-Day HA, also calculated for the 10-kg child, assumes a limited exposure period of one to two weeks. It is generally derived from a study of less than 30-days' duration.

Longer-Term Health Advisory. Longer-Term HAs are derived for both the 10-kg child and the 70-kg adult and assumes a human exposure period of approximately seven years (or 10% of an individual's lifetime). The Longer-Term HA is generally derived from a study of subchronic duration (exposure for 10% of an animal's lifetime, which is approximately 90 days to one year depending on the animal species).

Lifetime Health Advisory. The Lifetime HA is derived for the 70-kg adult and assumes an exposure period over a lifetime (approximately 70 years). The Lifetime HA is generally derived from a study of chronic duration (approximately two years in rodent species), but subchronic studies may be used by adjusting the uncertainty factor employed in the calculation.

Route of Administration

In all cases, the route of choice is oral exposure with the vehicle, in decreasing order of preference, being drinking water, gavage, and diet. Inhalation, subcutaneous, or intraperitoneal administration data are used on a case-by-case when no oral or other satisfactory data is available.

Test Species

The preferred species for assessing health effects is humans. However, as data in humans does not usually provide reliable dose-response information, selection of an appropriate animal model is usually required. The animal model selection is based upon its similarity to man in its pharmacokinetics handling of the chemical under evaluation. When different animal models vary considerably in their response to the chemical, the most sensitive, relevant species is selected. However, depending upon the toxicity of the chemical and the scope of the data available, information from all three sources may be used.

Assumptions Used in a Health Advisory

The HA values are presented under the Quantification of Toxicological Effects heading of the document and are based upon the assumptions listed in Table 3.

For consistency in calculation, the protected individual is considered to be either a 10-kg child, the individual likely to be most adversely

Table 3. Standard Assumptions Used to Develop Health Advisories

Protected individual

- One-day HA: 10-kg child
- Ten-day HA: 10-kg child
- Longer-term: 10-kg child and 70-kg adult
- Lifetime HA: 70-kg adult
- Cancer risk estimates: 70-kg adult

Volume of drinking water ingested/day

- 10-kg child: 1 liter
- 70-kg adult: 2 liters

Relative source contribution

- In absence of chemical-specific data:
  - 20% for organics and inorganics

Uncertainty factors

- 10:
  - NOAEL from human study
- 100:
  - LOAEL from human study
  - NOAEL from animal study
- 1,000:
  - LOAEL from animal study
  - NOAEL from animal study of less than lifetime duration (when calculating Lifetime HA)

Note: HA = Health Advisory; LOAEL = lowest-observed-adverse-effect level; NOAEL = no-observed-adverse-effect level.

affected during short-term exposure periods, or the 70-kg adult, the assumed average weight of an adult human exposed for periods from one year to a lifetime. It is further assumed that the average drinking water intake for a child and an adult are 1 and 2 liters per day, respectively. Additionally, if actual exposure data is not available, it is assumed that drinking water accounts for 20% of a person's total intake of organic and inorganic chemicals. This final assumption is used only when calculating the Lifetime HA when exposures from other sources (e.g., air or food) may become significant in the risk assessment.

Standard uncertainty factors (UFs) are also assumed during the HA calculation. These are based upon the type of data available. If the study was conducted in humans and a no-observed-adverse-effect level (NOAEL) is defined, a UF of 10 is used. If only the lowest-observed-adverse-effect level (LOAEL) has been established in humans, an additional factor of 10 is added and a UF of 100 is employed. If a NOAEL from an animal study is used to derive the HA, a UF of 100 is typically used. For HAs based upon animal data with only a LOAEL, another factor of 10 is used, giving a UF of 1000. The UF of 1000 may also be used when a Lifetime HA is based upon a NOAEL from an animal study of less-than lifetime's duration, (e.g., increased uncertainty associated with using shorter-term data to predict longer-term effects). Finally, it is emphasized that the selection of UFs requires case-by-case scientific judgments. Thus, deviations from these basic guidelines may be required when considering the total database for a specific chemical.

Calculation of Health Advisories

As previously stated, HAs are based upon identification of the adverse health effects associated with the most sensitive and meaningful noncarcinogenic endpoint of toxicity. The induction of this effect is related to both a particular exposure level and a specific period of exposure and is most often determined from the results of experimental animal studies. The general formula used to calculate HA values is as follows:

$$\text{HA} = \frac{(\text{NOAEL or LOAEL}) \times (\text{BW})}{(\text{UF}) \times (\_\ \text{L/day})} = \text{mg/L}\ (\_\ \mu\text{g/L})$$

where NOAEL or LOAEL = no or lowest-observed-adverse-effect level in mg/kg bw/day)

BW = assumed body weight of a child (10 kg) or an adult (70 kg)

UF = uncertainty factor (10, 100, or 1000)
_ L/day = assumed daily water consumption of a child (1 L/day) or an adult (2 L/day).

If the available data is derived from inhalation studies, the total absorbed dose (TAD) must first be determined before calculating the HAs. This is accomplished by converting the NOAEL or LOAEL, expressed in mg/m^3, to mg/kg bw/day by factoring in the respiratory rate for humans, 1 m^3/hr (if data derives from animal studies), the exposure duration, and the fraction of the test substance assumed to be absorbed. These factors will vary with the individual study protocols and chemicals studied.

One-Day and Ten-Day Health Advisories

The preceding formula is used for the One-Day and Ten-Day HAs by inserting the data for a 10-kg child consuming 1 liter of water per day, the appropriate UF and the NOAEL or LOAEL derived from a study of appropriate duration (e.g., one to five successive daily doses for the One-Day HA and 7 to 30 successive daily doses for the Ten-Day HA).

Longer-Term Health Advisories

Two values are calculated for the Longer-Term HA, using data for both the 10-kg child consuming 1 liter per day and the 70-kg adult consuming 2 liters per day along with the NOAEL or LOAEL from the study of appropriate duration. In this case, a 90-day to one-year animal study representing approximately 10% of an individual's lifetime and the appropriate UF for the type of data available are employed.

Lifetime Health Advisory

The Lifetime HA represents that portion of an individual's total lifetime exposure to the chemical which is attributed only to drinking water. All other HA values are calculated based on the assumption that drinking water is the sole source of the contaminant. The Lifetime HA is derived in a three-step process, with the first two steps being mathematically equivalent to the procedure used for all other HA calculations. The third step in the calculation is added to factor in the relative contribution from other exposure sources of the chemical.

Reference Dose. Step 1 determines the reference dose (RfD), formerly called the acceptable daily intake (ADI). The RfD is an estimate of the daily exposure to the human population that is likely to be without appreciable risk of deleterious effects over a lifetime. It is derived from the NOAEL (or LOAEL), identified from a chronic (or subchronic) study, that is divided by the appropriate uncertainty factor(s).

Drinking Water Equivalent Level. From the RfD, a drinking water equivalent level (DWEL) is calculated in Step 2. A DWEL is defined as a medium-specific exposure level (e.g., mg/L in drinking water), assuming 100% exposure from that medium, which is considered to be protective for noncarcinogenic health effects over a lifetime of exposure. The DWEL is derived by multiplication of the RfD by the assumed body weight of an adult (70 kg) and then divided by the assumed daily water consumption of an adult (2 L/day). For drinking water the DWEL is expressed in mg/L or μg/L. If the contaminant is classified as a Group A or B carcinogen, the calculation is halted at this point. The Lifetime HA (Step 3 below) is not calculated, and the DWEL is given to provide the risk manager with a reference point to evaluate noncarcinogenic endpoints. This infers that carcinogenicity should be considered the toxic effect of greatest concern when lifetime exposure is anticipated.

Relative Source Contribution (RSC). For noncarcinogenic chemicals, the Lifetime HA is determined in Step 3 by factoring in other sources of human exposure to the chemical (e.g., air, food, etc.). Preferably, the relative source contribution (RSC) from drinking water is based on actual exposure data. If data are not available, a value of 20% is assumed for organic and inorganic chemicals.

Evaluation of Carcinogenic Potential

If the chemical is determined to be a known or probable human carcinogen, Lifetime HAs are not recommended. Rather, carcinogenic risk estimates are derived by employing a cancer potency (unit risk) value together with assumptions for lifetime exposure and the consumption of drinking water. The cancer unit risk is usually derived from the linear multistage model's 95% upper confidence limits. This provides a low-dose estimate of cancer risk to humans that is considered unlikely to pose a carcinogenic risk in excess of the stated values. Excess cancer risk estimates can also be calculated using the one-hit, Weibull, logit, and probit models. There is only limited understanding of the biological mechanisms involved in cancer to suggest that any one of these models is able to predict risk more accurately than another. Because each model is based upon different assumptions, the estimates that are derived can differ by several orders of magnitude.

HEALTH ADVISORY DEVELOPMENT STATUS

Completed Health Advisories

Health Advisories for the chemicals listed in Table 4 have been completed and are available for use by any interested organization or individual.

National Pesticide Survey

The ODW has entered into a joint venture with EPA's Office of Pesticide Programs (OPP) to monitor those pesticides either known to have occurred in drinking water or are most likely to be found in groundwater. This joint venture is known as the National Pesticide Survey (NPS). An important element of the overall NPS is the development of HAs for all pesticides anticipated to be detected in water samples. This will allow the NPS manager to issue immediate health guidance when any pesticides are

Table 4. Completed ODW Health Advisories

Acrylamide	Endrin
Alchlor	Epichlorohydrin
Aldicarb	Ethylbenzene
Arsenic	Ethylene glycol
Barium	Heptachlor
Benzene	Hexachlorobenzene
Cadmium	*n*-Hexane
Carbofuran	Lead
Carbon tetrachloride	Lindane
Chlordane	Mercury
Chlorobenzene	Methoxychlor
Chromium	Methyl ethyl ketone
Cyanide	Nickel
2,4-D	Nitrate/Nitrite
1,2-Dibromo-3-chloropropane (DBCP)	PCBs
m/o-Dichlorobenzene	Pentachlorophenol
1,2-Dichloroethane	Styrene
1,1-Dichloroethylene	Tetrachloroethylene
cis-1,2-Dichloroethylene	Toluene
trans-1,2-Dichloroethylene	Toxaphene
Dichloromethane	2,4,5-TP
Dichloropropane	1,1,1-Trichloroethane
p-Dioxane	Trichloroethylene
Dioxin	Vinyl chloride
Ethylene dibromide (EDB)	Xylenes
	Legionella

Note: ODW = Office of Drinking Water.

discovered in drinking water supplies. Thus, an early step in the NPS was to compile a list of chemicals to be evaluated during the sampling and analysis effort and for which HAs were needed. This list of chemicals was complied based upon usage, water solubility, persistence in soil, and soil-water adsorption partition coefficient information. The list of chemicals resulting from this analysis is shown in Table 5. The HAs for these pesticides have been drafted and are currently in the review and revision process discussed earlier. Completion of these HAs is scheduled to coincide with initiation of full-scale groundwater sampling efforts under the NPS during FY 88.

Table 5. Proposed List of Analytes for the National Pesticide Survey

Acifluorfen	Diuron
Alachlor	Ethylene dibromide (EDB)
Aldicarb	ETU/EDBCs
Ametryn	Endothall
Ammonium Sulfamate	Fenamiphos
Atrazine	Fluometuron
Baygon	Fonofos
Bentazon	Glyphosate
Bromacil	Hexazinone
Butylate	Maleic hydrazide
Carbaryl	MCPA
Carbofuran	Methomyl
Carboxin	Methyl parathion
Chloramben	Metolachlor
Chlordane	Metribuzin
Chlorothalonil	Oxamyl
Cyanazine	Paraquat
Cycloate	PCP
Dalapon	Picloram
1,2-Dibromo-3-chloropropane (DBCP)	Prometone
DCPA/Dacthal	Pronamide
Diazinon	Propachlor
Dicamba	Propazine
2,4-D	Propham
1,2-Dichloropropane	Simazine
Dieldrin	Trifluoralin
Dimethrin	2,4,5-T
Dinoseb	2,4,5-TP
Diphenamid	Tebuthiuron
Disulfoton	Terbacil
	Terbufos

Other aspects of the NPS monitoring program already completed or nearing conclusion include development of analytical methods, selection of a hydrogeology scheme, finalization of sampling technique, and a pilot sampling survey. This survey will ultimately involve approximately 1500 groundwater wells, weighted toward areas of probable occurrence as influenced by pesticide usage and hydrogeology data.

Unregulated Volatile Organic Chemicals

Section 1445 of the Safe Drinking Water Act (SDWA) directs the EPA to require public drinking water systems to monitor for unregulated volatile organic chemicals (VOCs). These are VOCs for which there are no current primary drinking water regulations specifying a maximum contaminant level (MCL) or no requirement for a treatment technique. Monitoring for these chemicals will help EPA to determine the need for future regulation of these VOCs. An additional factor that influences potential regulation is the degree of toxicity for each VOC. To define this and to assist those faced with immediate VOC drinking water contamination problems, the ODW is preparing HAs for each of the chemicals listed in Table 6. For those VOCs that have adequate toxicity data, draft HAs are currently in preparation and are anticipated to be completed during FY 1988. For many of these VOCs, however, toxicity data is quite limited, which may prevent or delay the preparation of complete HA documents.

Department of the Army Munitions

The EPA has entered into a memorandum of understanding with the Department of Army to provide support in the preparation of HAs on various munition chemicals having the potential to contaminate drinking waters during their production, use, or disposal. Under this agreement, the Army provides the EPA with a priority ranking of those chemicals for which HAs are needed along with all relevant data the Army has assembled or developed on each munitions chemical. In addition, the Army remains the central point of contact to coordinate activities required under the agreement to include arranging for necessary visits by EPA personnel to Army facilities and other required support efforts.

The EPA's role under the memorandum of understanding is to assemble a team of health scientists to work with the Army to evaluate the available data, develop HAs when the data is sufficient or becomes available, define specific data deficiencies and/or problem areas encountered in the HA development process, and make recommendations for future database development. Table 7 lists the munition chemicals cur-

Table 6. List of Unregulated Volatile Organic Chemicals for Monitoring Under Section 1445 of the Safe Drinking Water Act

Monitoring Required for All Systems	Monitoring Required for Vulnerable Systems	Monitoring Required at the State's Discretion
Bromobenzene	Ethylene dibromide (EDB)	Bromochloromethane
Bromodichloromethane	1,2-Dibromo-3-chloropropane (DBCP)	*n*-Butylbenzene
Bromoform		Dichlorodifluoromethane
Bromomethane		Fluorotrichloromethane
Chlorobenzene		Hexachlorobutadiene
Chlorodibromomethane		Isopropylbenzene
Chloroethane		*p*-Isopropyltoluene
Chloroform		Napththalene
Chloromethane		*n*-Propylbenzene
o-Chlorotoluene		*sec*-Butylbenzene
p-Chlorotoluene		*tert*-Butylbenzene
Dibromomethane		1,2,3-Trichlorobenzene
m-Dichlorobenzene		1,2,4-Trichlorobenzene
o-Dichlorobenzene		1,2,4-Trimethylbenzene
trans-1,2-Dichloroethylene		1,3,5-Trimethylbenzene
cis-1,2-Dichloroethylene		
Dichloromethane		
1,1-Dichloroethane		
1,1-Dichloropropene		
1,2-Dichloropropane		
1,3-Dichloropropane		
1,3-Dichloropropene		
2,2-Dichloropropane		
Ethylbenzene		
Styrene		
1,1,2-Trichloroethane		
1,1,1,2-Tetrachloroethane		
1,1,2,2-Tetrachloroethane		
Tetrachloroethylene		
1,2,3-Trichloropropane		
Toluene		
p-Xylene		
o-Xylene		
m-Xylene		

Table 7. Army Munition Chemicals Scheduled for Health Advisory Development

Chemical
Trinitroglycerol (TNG)
Nitrocellulose (NC)
Trinitrotoluene (TNT)
Cyclotrimethylenetrinitramine (1-hexahydro-1,3–5-trinitro-1,3,5-triazine)(RDX)
Cyclotetramethylenetetranitramine (octahydro-1,3,5,7-tetranito-1,3,5,7-tetrazoline)(HMX)

rently identified for HA development. The HAs for trinitroglycerol and nitrocellulose have been completed, with the others in various stages of the preparation and review process.

In addition to these HAs, ODW has prepared toxicity profiles for the additional munition chemicals listed in Table 8. These chemicals are largely contaminants in and/or by-products of munitions manufacturing or waste disposal processes and may or may not be considered for future HAs. The toxicity profiles provide a brief survey of the properties of the chemical and the status of the toxicity database as is available from the published literature.

OTHER FACETS OF THE HEALTH ADVISORY PROGRAM

Federal-State Toxicology and Regulatory Alliance Committee

The Federal-State Toxicology and Regulatory Alliance Committee (FSTRAC) is a working group composed of EPA and state experts in the areas of risk assessment and risk management for drinking water contaminants. Goals of the committee, which meets approximately twice yearly, are to foster an attitude of cooperation and consistency in providing an exchange of information on setting drinking water standards between federal and state agencies. Some activities of the FSTRAC meetings include coordinating and updating the status of many EPA programs such as ODW Drinking Water Regulations, HAs, NPS, risk assessment guidelines, and Toxic Substance Disease Registry. Additionally, FSTRAC provides an opportunity for states to discuss their individual regulatory activities as well as methodology status, survey progress, and research activities and priorities.

Performance Improvement Project

As one of the EPA's Performance Improvement Projects (PIP), the ODW has conducted a series of workshops in all EPA regions on assess-

Table 8. List of Chemicals for Which Toxicity Profiles Have Been Prepared for the Department of the Army

1-Nitronaphthalene	2,5-Dinitrotoluene
1-Methyl-2-nitrobenzene	2,6-Dinitrotoluene
3,4-Dinitrotoluene	1-Methyl–4-nitrobenzene
3,5-Dinitrotoluene	1-Chloro-4-nitrobenzene
2,3-Dinitrotoluene	1,2-Dichloro-4-nitrobenzene

ing and managing drinking water contamination. The workshops are led by scientists and regulatory officials directly involved in the implementation of EPA's drinking water programs. The workshops, which are conducted over a period of two to three days each, stress the qualitative and quantitative risk assessment process. Additionally, presentations on the principles of pharmacokinetics, risk assessment, carcinogenicity, and toxicology are provided for the various classes of drinking water contaminants (i.e., inorganics, synthetic organics, and pesticides). The ODW PIP workshops focus primarily on the HA Program, its development, philosophy, and methodology. Analytical technology and treatment techniques are discussed as well as the communication of potential or existing health risks to the general public. Actual risk management case studies are presented to provide hands-on experience to the attendees for specific drinking water contaminants.

Emergency Response Network

The Emergency Response Network is a long established and very important component of ODW's HA Program. It is designed to provide state, local, and other concerned parties rapid access to existing information on drinking water contaminants. This service is provided through a systematic access to EPA experts, databases, HAs, Criteria Documents, and other regulatory documents. Figure 2 illustrates the processing of incoming requests for assistance. In brief, requests received by letter or telephone from the concerned party (regional and state EPA offices, state and local health departments, local water treatment facilities, or other concerned individuals or organizations) are logged in, classified, and referred to a specific chemical manager within the ODW Health Effects Branch. This staff member has ready access to other staff scientists, HAs and Criteria Documents, contractor support, and other national experts to formulate a response to the request. Depending upon the nature of the request and the degree of urgency, the response may be relayed to the requesting party via letter, telephone or conference call.

SUMMARY

In summary, the ODW Program is a multifaceted information clearinghouse activity that provides other government organizations and the public nonregulatory guidance on drinking water contaminants. Elements of the program range from extensive literature evaluations and assessments (e.g., preparation of HA documents) to informal, fast-response advice (e.g., responses under the Emergency Response Network). Although not a formal component of ODW's regulatory pro-

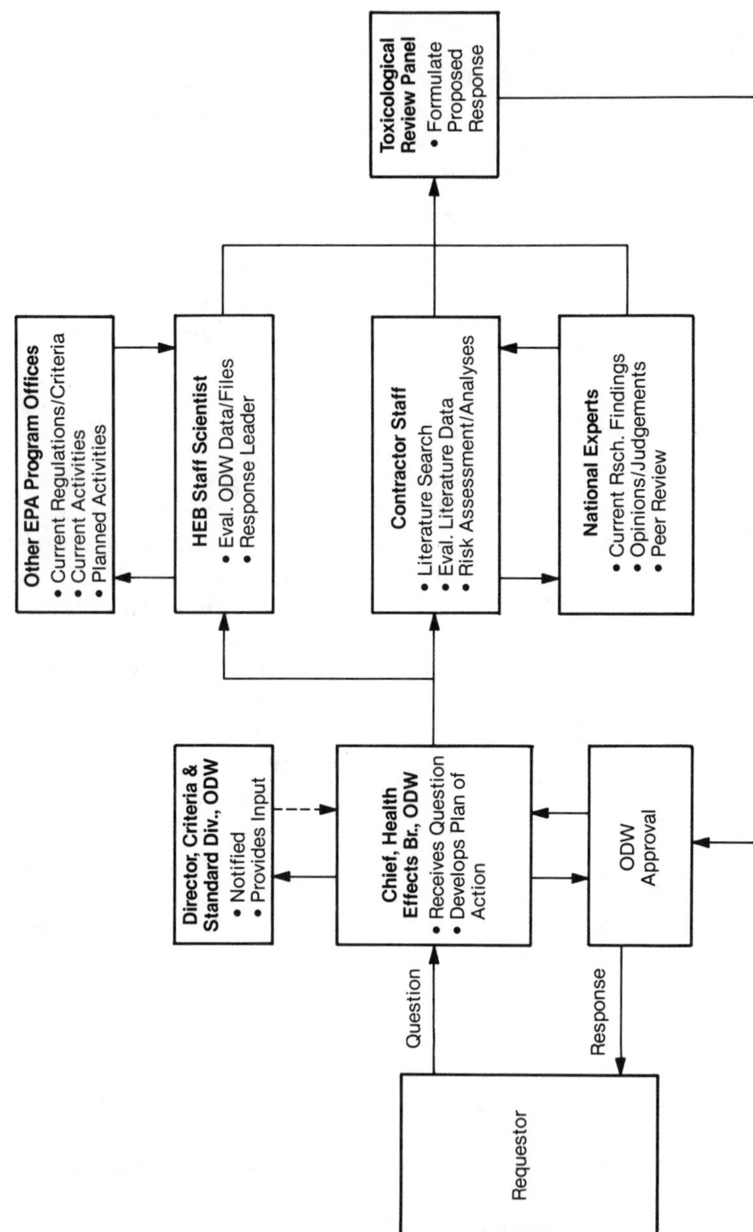

Figure 2. Procedure for processing requests under ODW's Emergency Response Network.

gram, the information assembled under the HA Program has a major impact on the identification of contaminants that should be given priority for future regulation. Thus, the HA Program constitutes a vital role in ODW's efforts to meet the regulatory requirements of the SDWA. The HA Program also serves as ODW's primary technology transfer mechanisms to health officials and the public on the health effects of drinking water contaminants.

CHAPTER 11

Compliance by Public Water Supply Systems with National Primary Drinking Water Regulations

Peter C. Karalekas, Jr., and John R. Trax

INTRODUCTION

Since the passage of the original Safe Drinking Water Act (SDWA) in 1974, the U.S. Environmental Protection Agency (EPA) has tracked the compliance by public water supply systems with the National Primary Drinking Water Regulations (NPDWR). EPA receives reports from the states with primary enforcement responsibility of all violations of the NPDWR and these violations are incorporated into the Federal Reporting Data System (FRDS). Of the 57 states and territories that are included in the public water systems supervision program under the SDWA, 54 have been granted primary enforcement responsibility, with 3 programs being administered directly by EPA regional offices.

In addition to information on violations, quarterly reports are submitted to EPA on the basic characteristics of public water systems (PWS) such as population served, source of supply, and treatment, which is called the inventory of public water systems and is also incorporated into FRDS. The following is the latest information in FRDS on both the characteristics of the systems and their compliance.

THE UNIVERSE OF PUBLIC WATER SYSTEMS

A public water supply system, which is subject to the provisions of the SDWA, is defined in the law and EPA regulations as a system that has at least 15 service connections or regularly serves an average of at least 25 individuals daily at least 60 days out of the year.

For purposes of regulation, public water systems are divided into three distinct groups. A *community water system* is a public system that serves year-round residents. A *noncommunity water system* is a public water system that serves nonresidential areas or customers such as restaurants, motels, and campgrounds. A *nontransient noncommunity water system* is a noncommunity system that regularly serves the same population, such as a school or factory.

Figure 1 shows the distribution of all public water systems in the United States and territories. At the end of fiscal year 1986 (i.e., September 30, 1986), there were a total of 202,459 systems of which 58,557 were community systems and 143,902 were noncommunity. Nontransient noncommunity are not broken out as a group and are included in the total of noncommunity systems. Of the community systems, surface water is the primary source of supply (19%) and serves 147 million people, which is 67% of the total population. Groundwater is used by 81% of the systems but supplies only 33% of the population. In the noncommunity water systems, 97% are served by groundwater.

For purposes of compliance analysis, community water systems are divided into five size categories, shown in Table 1. Table 1 also shows the distribution of the 58,557 community water systems, the largest number of which fall into the *very small* category with 37,334 systems; this group serves only 2.5% of the population. The *very large* systems form the smallest group, with 279 systems, but serve over 43% of the population.

COMPLIANCE WITH FEDERAL REGULATIONS

The following discussion of violations will pertain only to community water systems (CWS), since nationwide data are incomplete on noncommunity systems. For oversight and management purposes, water systems in which violations occur are divided into two categories. Significant noncompliers (SNCs) are those systems which have more serious and more frequent violations. They are defined for each contaminant category in Table 2. Minor noncompliers are systems which have at least one violation but were not serious enough to be classified as an SNC.

Figure 2 illustrates compliance with the microbiological, turbidity, and total trihalomethane regulations. During FY 86, 71% or 41,776 of the

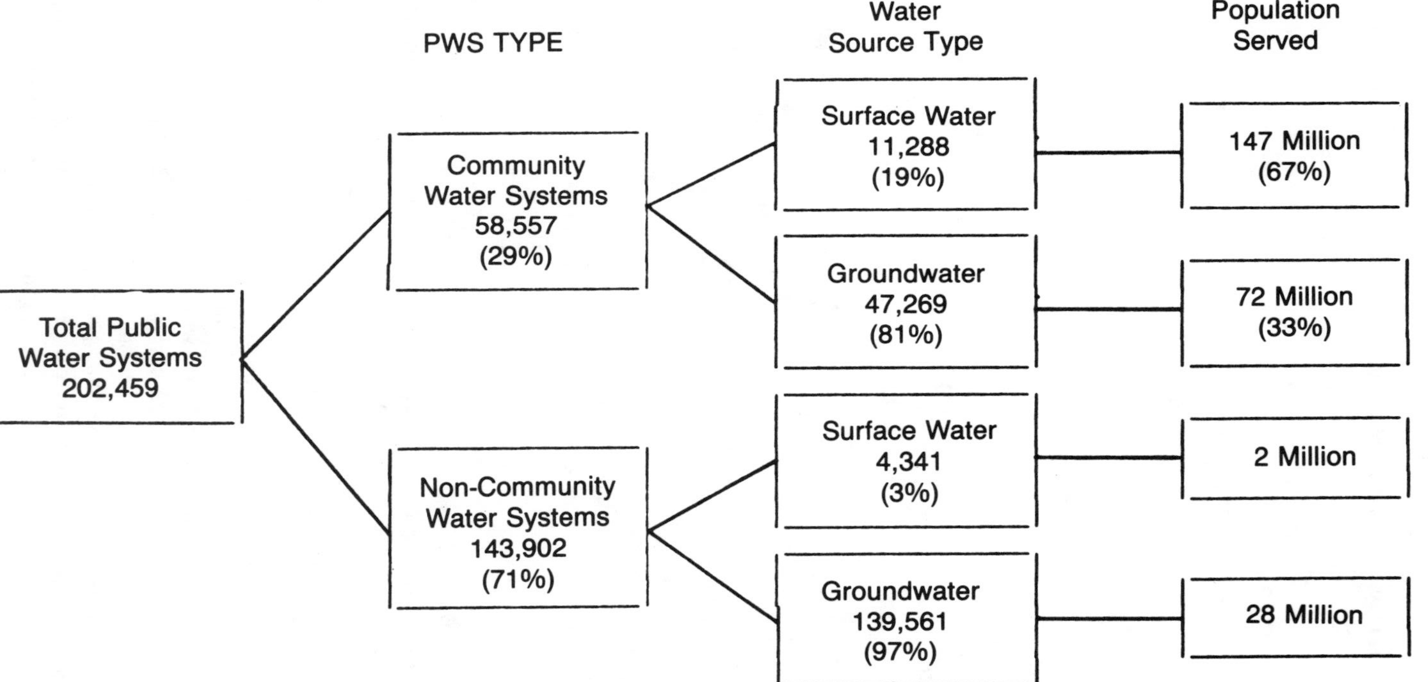

Figure 1. Distribution of public water systems by system type and source water, FY 1986.

Table 1. Distribution of Community Water Systems by Size of Population Served, FY 1986

System Size (population served)	No. of Systems	Percentage of Systems	Population Served (millions)	Percentage of Population Served
Very small (≤500)	37,334	63.8	5.5	2.5
Small (501–3,300)	13,999	23.9	19.2	8.8
Medium (3,301–10,000)	4,121	7.0	23.5	10.7
Large 10,001–100,000)	2,824	4.8	76.3	34.8
Very large (>100,000)	279	0.5	94.5	43.2
Total	58,557	100	219	100

total 58,537 community water systems had no reported violations. Minor noncompliers comprised 27%, and SNCs 2%. Table 3 shows the distribution of the violations by system size. It can be seen that the great majority of SNCs (83%) and minor noncompliers (72.9%) occur in the very small system category. In the very large category, with populations greater than 100,000, there are only 4 systems classified as SNCs and 30 with minor violations.

Table 4 shows the distribution of SNCs by contaminant category. Violations of the microbiological (i.e., coliform) maximum contaminant level (MCL) constitute the largest group of systems with 361 sytems in this category. Microbiological monitoring and reporting violations (micro-M/R) are the next largest individual category, with 204 systems.

The next group of SNCs are those with violations of either a chemical (other than TTHM) or radiological (chem/rad) MCL. Figure 3 illustrates the very high level of compliance with chem/rad MCLs. Of the national total of 58,557 community water systems, 56,544 (96.6%) were in total compliance. Minor noncompliers comprised 2.3%, and SNCs only 1.1%.

Table 5 lists the contaminants that were regulated during FY 1986 and the number of SNCs for each contaminant. The most frequently violated contaminants were combined radium, with 164 SNCs, and fluoride, with 159. The least frequently violated contaminant was 2,4-D, with only 3 SNCs. Of particular note are the 30 SNCs for lead. Although there has been widespread use of lead-containing materials such as lead pipe, lead solder, brass, and bronze in water distributions systems, lead violations are rare. This may be explained by current monitoring requirements, for

Table 2. Definition of a Significant Noncomplier, FY 1986[a]

Group I

1. violates the microbiological MCL for 4 or more months during any 12-consecutive-month period (called *microbiological MCL*)
2. violates the turbidity MCL for 4 or more months during any 12-consecutive-month period (called *turbidity MCL*)
3. takes no microbiological samples during any 12-consecutive-month period (called *microbiological M/R*)
4. takes no turbidity samples during any 12-consecutive-month period (called *turbidity M/R*)
5. takes no TTHM samples during any 12-consecutive-month period (called *TTHM M/R*)
6. takes some but not all of the required microbiological samples during any 12-consecutive-month period, but the samples taken violate the MCL (called *microbiological aggregate*)
7. takes some but not all of the required turbidity samples during any 12-consecutive-month period, but the samples taken violate the MCL (called *turbidity aggregate*)

Group II

8. has a chemical (other than TTHM) or radiological violation, during the most recent compliance period, which not only exceeds the MCL but exceeds the level above which variances or exemptions may not be granted (called *chem/rad contaminant concentration level*)
9. exceeds the MCL for TTHM for 2 or more annual averages during the federal fiscal year (called *TTHM MCL*)

Group III

10. fails to monitor for, or report the results of, any one of the currently regulated inorganic, organic (other than TTHM), or radiological contaminants since the federal requirements for the contaminant became effective—June 24, 1977 (called *chem/rad M/R*)

Note: chem/rad = chemical or radiological; MCL = maximum contaminant level; M/R = monitoring and reporting; SNC = significant noncomplier; TTHM = total trihalomethanes.

[a] A community water system is an SNC if it meets *any* of the ten criteria.

lead which involve taking a single sample per year for a surface water supply and one sample every three years for a groundwater supply. These monitoring requirements were not designed to detect the wide variation one would expect to find at consumer's taps as a result of lead corrosion. Consequently, these violations will either represent lead in the source of supply or the chance finding of lead at a tap in the distribution system. More rigorous monitoring would undoubtedly find more violations.

The final area of compliance deals with monitoring and reporting (M/R) for chem/rads. Figure 4 shows the very high level of compliance, with 97.1% systems having conducted at least one analysis for chem/rads and only 2.9% classified as SNCs, which indicated no monitoring had been done for chem/rads.

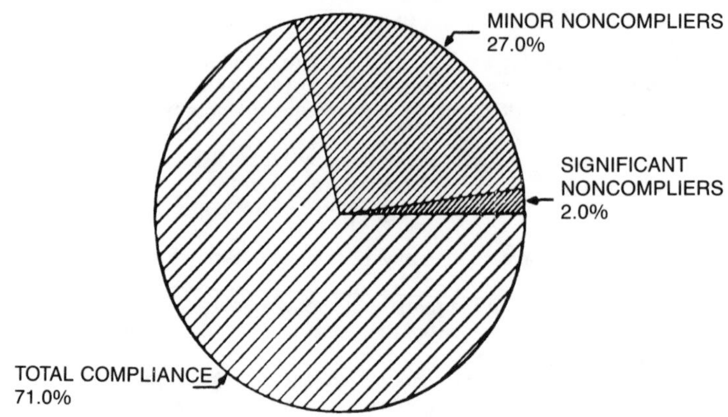

Figure 2. National compliance/violation status–microbiological, turbidity, and total trihalomethanes–FY 1986.

In summary, these data indicate a very high level of compliance by most categories of water systems. The so-called small system problem is evident, with most violations occurring in the very small systems. Despite the efforts to focus attention and resources on these small systems, this problem is likely to persist as new regulations are promulgated.

Table 3. Distribution of Micro/Turb/TTHM Violations (Group 1) by Size of Population Served, FY 1986

System Size (Population Served)	SNC		Minor Noncomplier	
	Number	Percent	Number	Percent
Very small <500	1033	83	11,341	72.9
Small 501–3,300	152	12	2,936	18.9
Medium 3,301–10,000	25	2	728	4.7
Large 10,001–100,000	26	2	506	3.3
Very large >100,000	4	0.3	30	0.2
Total	1,240	100[a]	15,541	100.0

Note: micro/turb/TTHM = microbiological, turbidity, and total trihalomethane; SNC = significant noncomplier.

[a] Rounded.

Table 4. Distribution of SNCs (Microbiological, Turbidity, Total Trihalomethanes) by Contaminant Category, FY 1986

Contaminant Category	SNC Criteria Number	No. of SNC	Percentage of SNC
Micro-MCL	1	361	29
Turb-MCL	2	151	12
Micro-M/R	3	204	16
Turb-M/R	4	97	8
TTHM-M/R	5	3	<1
Micro-Aggregate	6	13	1
Turb-Aggregate	7	4	<1
Micro-MCL & Turb-MCL	1 & 2	170	14
Micro-M/R & Turb-M/R	3 & 4	230	19
Micro-MCL & Turb-M/R	1 & 4	5	<1
Micro-Aggregate & Turb-M/R	4 & 6	1	—
Micro M/R & Turb-Aggregate	3 & 7	1	—
Total SNCs	1 – 7	1240	100

Note: MCL = maximum contaminant level; M/R = monitoring and reporting; SNC = significant noncomplier; TTHM = total trihalomethanes.

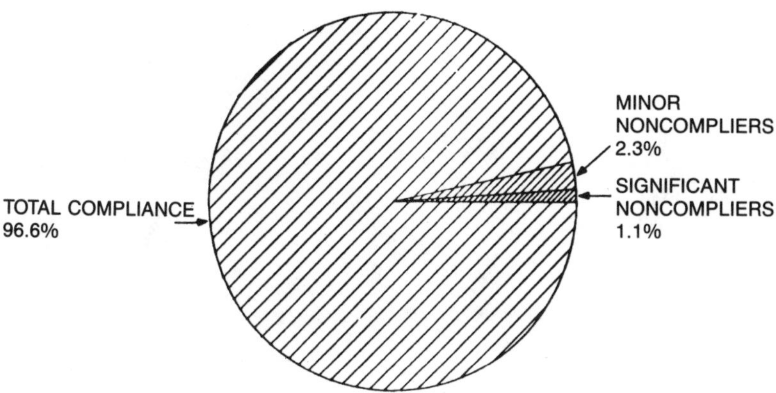

Figure 3. National compliance/violation status, chem/rad contaminant concentration level SNCs, FY 1986.

Table 5. Distribution of Chem/Rad Contaminant Concentration Level Violations by Contaminant Category, FY 1986

Contaminant	No. of SNC	No. of Minor Noncomplier	Total No. Systems in Violation
Arsenic	24	20	44
Barium	24	15	39
Cadmium	10	6	16
Chromium	6	3	9
Fluoride	159	581	740
Lead	30	0	30
Mercury	23	0	23
Nitrate	141	356	497
Selenium	35	95	130
Silver	15	0	15
Endrin	11	2	13
Lindane	6	0	6
Methoxychlor	6	0	6
Toxaphene	6	0	6
2,4-D	3	0	3
2,4,5-TP (Silvex)	6	0	6
Total TTHM	14	0	14
Gross alpha excluding radon and uranium	69	165	234
Gross alpha including radon and uranium	0	6	6
Combined radium	164	235	399
Total PWS	652[a]	1361[a]	2013[a]

Note: PWS = public water systems.

[a] Some systems are violators of more than one contaminant.

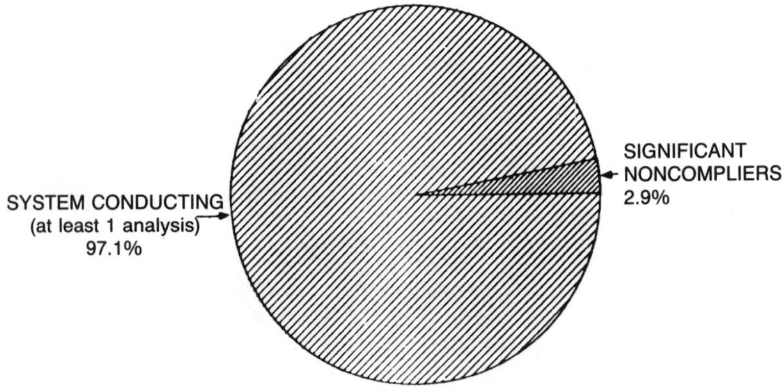

Figure 4. National compliance/violation status, chem/rad monitoring/reporting SNCs, FY 1986.

CHAPTER 12

Drinking Water Quality and Water Treatment Practices: Charting the Future

Edward J. Calabrese and Charles E. Gilbert

INTRODUCTION

The quality of community drinking water is affected by the source of the water, the type and condition of the distribution system, the treatment of the water, the capacity of the water utility to effectively manage the system, and other factors. In the United States, approximately 10% of the water systems are large and serve about 90% of the population, while 90% of the systems are small and serve 10% of the population. More specifically, 88.4% of the nation's 59,000 drinking water systems serve less than 3300 people, while about 38,500 systems serve less than 500 people.[1] Most of the smaller systems pump their supplies from groundwater, possibly through a chlorinator and then to a simple distribution system, all supervised by an untrained part-time operator. On the other hand, many of the large systems in the United States use surface waters of varying quality and have complex treatment and distribution systems monitored and serviced by a highly skilled professional staff. Other water supply systems fall somewhere in between these two extremes.

An estimated 1 to 2 billion gallons of drinking water are provided on a daily basis in the United States by public water systems. In order to comply with health and other standards for treating that quantity of water, nearly $300 million was spent on the purchase of chemicals by U.S. water utilities in 1982.[1] The specific chemicals and the quantities used by the water treatment industry in 1982 are listed in Table 1. Of greatest use were chlorine for disinfection; alum, ferric chloride, and ferric sulfate for coagulation; and calcium oxide, soda ash, hydrated lime, and sodium hydroxide for softening. In general, chlorine has been viewed by the water utility industry as economical, effective, and convenient in virtually eliminating the transmission of bacteriological and viral diseases from drinking water. The use of chlorine has markedly expanded the usability of the nation's water resources by permitting the use of water not considered pristine and protected. In fact, chlorine has been considered "virtually the only irreplaceable chemical in the water."[1]

Troubling questions, however, about the quality of our drinking water have been frequently raised over the past few years. Many of these questions have been scientifically based and have led large numbers in our population to not only question the quality of our water but to consciously search out alternative sources that they believe are better tasting, more aesthetic, and more healthful.[2] State public health officials are concerned as well. An April 1987 national survey of state public health officials revealed that drinking water contamination ranked first of 27 environmental health concerns.[3] There is widespread concern about the public health advantages and disadvantages of the most widely used drinking water treatment technologies. Of greatest relevance is the issue of disinfection.

Table 1. Chemicals in Water Treatment

Chemical	Quantity (1000 tons)	Chemical	Quantity (1000 tons)
Coagulation		Softening	
Alum	250.0	Calcium oxide	350.0
Ferric chloride	80.0	Hydrated lime	170.0
Ferric sulfate	100.0	Sodium dioxide	135.0
Polyelectrolytes	4.5	Carbon dioxide	18.0
		Soda ash	200.04
Disinfection		Miscellaneous	
Chlorine	500.0	Fluoride compounds	38.0
Hypochlorite	2.5	Activated carbon	10.0
Sodium chlorite	6.4	Phosphate	16.0
Ammonia	2.5	Sodium chloride	6.5
		Copper sulfate	1.5

Source: Manwaring.[1]

DISINFECTANTS AND THEIR BY-PRODUCTS

Disinfection has played an enormous role in the maintenance of a high standard of health in the United States. The consequences to society of *not* disinfecting would be so calamitous that it is beyond rationality to even consider eliminating it. However, the process of chlorination as practiced in the United States may lead to the formation of a variety of disinfectant by-products: chloroform; chloral, di-, and trichloroacetic acids; and several dozen others. A few of these by-products are known to be carcinogenic in animal models and may be present at levels in drinking water of up to and greater than 100 μg/L. In fact, the U.S. EPA has established a maximum contaminant level (MCL) for total trihalomethanes (TTHMs) of 100 μg/L in an effort to reduce exposure to these agents. Of concern to regulators is that these by-products have not been sufficiently characterized in toxicological terms for risk assessment.

The EPA and the waterworks industry have identified practical means to reduce significantly THM and other by-products while retaining the use of chlorine, and adopting the use of other disinfectants such as chloramine, chlorine dioxide, and ozone. Some of these adjustments in the treatment train have certain drawbacks: such as, with chloramine usage, a less effective biocide, possible concern with nuisance growths and odors, and more frequent cleaning of flocculation basins. Any switch to alternative disinfection processes must be predicated on the notion that a significant risk to infectious disease is unacceptable. While the margin of safety has been judged to be adequate to ensure the health of the public when treatment train/disinfectant techniques are modified, there is nonetheless a general paucity of data in this area to validate such professional judgments.

Microbial Aspects

In order to assess objectively the overall public health significance of the potential carcinogenic effects of disinfectants such as chlorine and their by-products, it is first necessary to consider chlorine as a disinfectant and its role in preventing waterborne disease.

Typhoid Fever

The public health importance of disinfection of drinking water with chlorination was convincingly demonstrated with the striking decrease in typhoid fever deaths in the first three to four decades of this century. Around 1900, approximately 25,000 deaths and considerably higher numbers of actual nonfatal cases of typhoid fever occurred annually in the United States. Forty percent of these typhoid fever cases were attributed to waterborne transmission. However, by the mid-1930s there was a

10-fold decrease in typhoid fever deaths, with much of the credit going to improved water disinfection. The first application of chlorine for municipal water disinfection occurred in Jersey City in 1908, and its widespread adaptation coincided with the decreased risk of typhoid fever.

A major key to the success of chlorination was that the chlorine level that was broadly acceptable from a taste and odor perspective was also effective against the fecal contamination indicator organism *Escherichia coli* as well as the bacterial pathogen *Salmonella typhosa,* the causative agent of typhoid fever. In those early years, to satisfactorily disinfect the supply and to avoid complaints of bad-tasting water due to residual chlorine, a common practice emerged to add about 0.25 ppm total chlorine, which commonly yielded less than 0.1 ppm *o*-toluidine-detectable residual after a 15-minute contact time.[4]

Viruses and Protozoa

The etiology of waterborne illness in the United States has shifted from diseases of bacterial origin to those caused by viruses and protozoa, agents more resistant to disinfection than pathogenic bacteria. By the 1950s, concern emerged over the capacity of chlorine to sufficiently inactivate human pathogenic viruses. Studies had revealed that enteric viruses were considerably more resistant to chlorine than bacterial indicators and bacterial pathogens. About the same time, it was found that the cysts of *Entamoeba histolytica,* the cause of amoebic dysentery, were highly resistant to chlorine. Fortunately, the amoebic cysts, because of their relatively large size, can be removed by filtration.[5]

To ensure virus inactivation, the intensity of chlorine treatment needed to increase. From 0.25 ppm total chlorine residual with 15 minutes' contact time in the early 1900s, the treatment practice was elevated to 0.4 to 1.0 ppm free residual chlorine for at least a 30-minute contact time.[6,7] However, since prechlorination was included in most conventional treatment processes, the actual chlorine contact time of drinking water was often 2–4 hours at high chlorine levels.

Despite increased chlorine treatment and conventional full treatment practice, recent research has uncovered surprising positive virus findings. Of concern is that viral disinfection models based on typical virus levels underpredicted actual measured values. Furthermore, limited human investigations have found that many enteric pathogens (responsible for gastroenteritis) are highly potent infectious agents for healthy human volunteers.[8,9] It is very likely that the high-risk segments of the population would become infected at a dose of microbes much lower than previously thought to be safe. According to Akin and Hoff,[5] microbial pathogens, "perhaps at levels undetectable in water with current methodology," may cause disease.

Since 1970, waterborne giardiasis* has emerged as a serious public health concern in this country. From 1970 to 1982, there were approximately 20,000 cases of waterborne giardiasis in the United States. Since this is believed to be a grossly underreported disease, the actual incidence is likely to be considerably higher. According to EPA,[10] the populations at higher risk for waterborne giardiasis live in New England, Pennsylvania, Colorado, California, and the Pacific Northwest. These areas traditionally use surface water that is generally free of human sewage and low in bacterial contaminants. Consequently, water treatment uses primarily disinfection, not filtration. According to EPA, 56 million people are potentially at risk for waterborne giardiasis as a result of unfiltered community water systems. Actually the number at risk is considerably greater, since that number does not include people served by noncommunity water systems, people served by facilities where filtration exists but the system has deficiencies, and seasonal consumers of untreated surface waters (e.g., hikers, campers, fishermen).

Since most giardiasis outbreaks (67%) and disease (52%) have been associated with contamination from either untreated surface water or surface water with disinfection as the only treatment, EPA attempted to use these figures to estimate the possible decrease in giardiasis outbreaks if all systems using surface water included filtration. If 8 outbreaks and 2700 cases of illness occurred annually, then 5.3 outbreaks and 1400 cases of illness could be prevented per year. These estimates are likely to grossly underestimate the actual benefit, since waterborne giardiasis is grossly underreported. The order of efficacy for *Giardia* disinfectants conventionally used is

$$\text{ozone} > ClO_2 > \text{free chlorine} > \text{chloramine}$$

Free chlorine is approximately 8 times more effective for *Giardia* cyst inactivation than chloramine.[11] Thus, where there is no filtration and a reasonable risk of *Giardia* contamination, chloramine should not be employed as a primary disinfectant.

Modified Treatment Practices

We know that full conventional treatment, involving prechlorination, flocculation, sedimentation, rapid filtration, and postdisinfection, is

*Giardiasis is an extremely unpleasant disease in most humans. The severity is highly variable with most acute disease symptoms lasting from one to four weeks. In general, the disease is characterized by diarrhea, fatigue, abdominal cramps, and nausea. *Giardia* cysts, like those of *Entamoeba histolytica,* can be removed by filtration.

effective in producing drinking water that will not spread infectious disease. According to Akin and Hoff,[5] there are no clinical or epidemiological data to refute its effectiveness. They assert that documented episodes of illness have resulted only when incomplete treatment or deficiencies in unit processes have occurred and that treatment lacking disinfection procedures may not be sufficient for health protection. Disease has occurred where limited treatment has not effectively dealt with unanticipated circumstances, resulting in documented waterborne outbreaks of gastroenteritis[12] and giardiasis.[5] Lippy and Waltrip[13] assessed the literature on 625 waterborne infectious disease outbreaks over the period of 1946–1980. These outbreaks occurred, for the most part, in small surface water and groundwater systems where full conventional treatment was not applied. Moreover, a majority of the outbreaks occurred in systems that failed either to apply or to maintain any treatment. We can thus conclude that the risk of waterborne infectious disease is enhanced in the absence of full conventional treatment.

Despite the apparent necessity of full conventional water treatment, the waterworks industry has generally accepted, at least in practice, that it is not needed for all water supplies. Numerous supplies have indeed operated this way for decades without the public's experiencing a waterborne disease outbreak. Many drinking water sources, especially groundwater, are not believed to harbor pathogenic microorganisms and are consequently believed to need no treatment. It is acceptable to apply limited disinfection to groundwater and still satisfy regulatory requirements.

The THM problem has had a tremendous impact on chlorination practices. Since previous animal toxicity studies have indicated that consumption of chlorine at levels of up to 200 mg per liter of drinking water has no adverse health effects,[14,15] Akin and Hoff[5] believe that application rates of 10 mg/L as a prechlorination step have been widespread. They feel that many utilities drawing water from rivers and streams have used amounts far in excess of the recommended 0.5 to 1.0 mg/L of residual chlorine for a 30-minute contact period. Now that the emphasis is to lower the THMs in the most economical way without affecting disinfection, many of these utilities have changed to the lower amount of chlorine and the shorter contact time.

How communities reduce THM levels has important implications for the safety of drinking water. A 1983 study by EPA revealed that in 24 communities where disinfection treatment processes were modified to lower THM levels, 9 had modified the point of chlorination, 10 had changed to chloramination, and 1 had switched to chlorine dioxide.

Point of Chlorination. A simple switch in the treatment train from a presedimentation chlorination to a postsedimentation chlorination has been successful in lowering THMs. However, the chlorine contact time for microbial destruction prior to distribution of the water may be markedly diminished in such situations.

Chloramination. Chloramination, the most widely used of alternative disinfectant practices, is easily implemented by adding ammonia in a ratio of 3:1 just before or just after adding chlorine to produce a combined-chlorine residual. The combined form has good stability, which ensures a chloramine residual throughout the distribution system. Chloramination has been widely criticized by some public health officials, however, since the combined chlorine-ammonia residuals are the least effective of the principal disinfectants used today. In fact, the recent National Academy of Sciences (NAS) report on disinfectants has recommended that chloramination *not* be used as a primary disinfectant, especially where potential viral and parasitic cyst contamination exist. Nevertheless, Akin and Hoff[5] indicate that "in practice, chloramination has produced finished water that meets microbiological standards, and no known water outbreaks have been attributed to an inadequacy of this disinfectant."

In fact, in most U.S. water suppliers employing this process, chloramine is not used as the primary disinfectant. Commonly, ammonia is added to water having a slight chlorine residual just prior to its entry into the distribution system. This results in the formation of monochloramine, which serves the dual purpose of providing a disinfectant residual throughout the distribution system and reducing THM formation.

Chlorine Dioxide and Ozonation. Other alternative disinfectants, such as ClO_2 and O_3, compare favorably with chlorine in their capacity for disinfection (Tables 2 and 3). In fact, chlorine dioxide is a stronger oxidant than chlorine and is known to be an effective biocide. However, it has been tested with comparatively few waterborne pathogens. Ozone can also be as effective as chlorine in primary disinfection of drinking water, but has the disadvantage, when used by itself, of not providing a residual in the water to protect the distribution system against subsequent contamination.[11]

What are the microbiological implications of these alterations in the disinfection treatment process? Akin and Hoff[5] contend that the effectiveness of the 0.5 mg/L chlorine for 30 minutes of contact before distribution to the consumer has not been sufficiently assessed under worst-case conditions. They also contend that the reduction in disinfection

Table 2. Summary of Major Possible Disinfection Methods for Drinking Water[a]

Disinfection Agent[c]	Technological Status	Efficacy in Demand-Free Systems[b]			Persistence of Residual in Distribution System
		Bacteria	Viruses	Protozoan Cysts	
Chlorine[d]	Widespread use in U.S. drinking water				Good
As hypochlorous acid (HOCl)		+ + + +	+ + + + +	+ +	
As hypochlorite ion (OCl^-)		+ + +	+ +	NDR[e]	
Ozone[d]	Widespread use in drinking water outside United States, particularly in France, Switzerland, and the province of Quebec	+ + + +	+ + + +	+ + + +	No residual possible
Chlorine dioxide[d]	Widespread use for disinfection (both primary and for distribution system residual) in Europe, limited use in United States to counteract taste and odor problems and to disinfect drinking water	+ + + +	+ + + +	NDR[e]	Fair to good (but possible health effects)
Iodine	No reports of large-scale use in drinking water				Good (but possible health effects)
As diatomic iodine (I_2)		+ + + +	+ + +	+ + +	
As hypoiodous acid (HOI)		+ + + +	+ + + +	+	
Bromine	Limited use for disinfection of drinking water	+ + + +[f]	+ + + +[f]	+ + +[f]	Fair
Chloramines	Limited present use on a large scale in U.S. drinking water	+ +	+	+	Excellent

Source: NAS.[11]

[a] Data from NRC (1980a), pp. 114–115.

[b] Ratings: + + + +, excellent biocidal activity: + + +, biocidal activity: + +, moderate biocidal activity: +, low biocidal activity: ±, of little or questionable value.

[c] The sequence in which these agents are listed does not constitute a ranking.

[d] By-product production and disinfection demand are reduced by removal of organics from raw water before disinfection.

[e] Either no data reported or only available data were not free from confounding factors, thus rendering them not comparable to other data.

[f] Poor in the presence of organic material.

Table 3. Status of Possible Methods of Drinking Water Disinfection[a]

Disinfection Agent	Suitability as Inactivating Agent	Limitations	Suitability for Drinking Water Disinfection[b]
Chlorine	Yes	Efficacy decreases with increasing pH; affected by ammonia or organic nitrogen	Yes
Ozone	Yes	On-site generation required; no residual; other disinfectant needed for residual	Yes
Chlorine dioxide	Yes	On-site generation required; interim MCL 1.0 mg/liter	Yes
Iodine	Yes	Biocidal activity sensitive to pH	No
Bromine	Yes	Lack of technological experience; activity may be pH sensitive	No
Chloramines	No	Mediocre bactericide; poor virucide	No[c]
Ferrate	Yes	Moderate bactericide; good virucide; residual unstable; lack of technological experience	No
High pH conditions	No	Poor biocide	No
Hydrogen peroxide	No	Poor biocide	No
Ionizing radiation	Yes	Lack of technological experience	No
Potassium permanganate	No	Poor biocide	No
Silver	No	Poor biocide; MCL 0.05 mg/liter	No
UV light	Yes	Adequate biocide; no residual; use limited by equipment maintenance considerations	No

Source: NAS.[11]

Note: MCL = maximum contaminant level.

[a] Data from NRC (1980a), p. 118.

[b] This evaluation relates solely to the suitability for controlling infectious disease transmission. See Conclusions.

[c] Chloramines may have use as a secondary disinfectant in the distribution system in view of their persistence.

potency by switching to chloramine as the primary disinfectant may permit low-level penetration of the more resistant pathogens. The prime microbiological concern associated with ClO_2 and O_3 is the likelihood that a sufficient residual disinfectant level will not be maintained, as a result of inexperience of workers with the process and of technological failure of unproven delivery systems. Thus, while both ClO_2 and O_3 have good potential as community disinfectants, there are very limited data in the United States upon which to judge their broad-based effectiveness.

Toxicological Aspects

Cardiovascular Disease

Research has demonstrated that consumption of drinking water with chlorine residuals within an order of magnitude of those found in community drinking water causes an increase in serum cholesterol[16] and produces indications of myocardial hypertrophy and arteriosclerosis in rabbits and pigeons[17] reared on a diet marginal in calcium. At normal levels of dietary calcium, the development of chlorine-induced cardiovascular alterations in these animal models is reduced by a factor of 3. Of concern to the public health community is that certain subgroups within the U.S. population consume diets deficient in calcium at the levels used in the animal experiments. For example, the median calcium intake by blacks aged 18 and older is below the recommended daily allowances (RDA), while the median intake for whites falls below the RDA by age 35.[18] More recently, limited human clinical studies have suggested that consumption of chlorinated water increases the serum cholesterol levels in a dose-dependent way in volunteers given a marginally deficient diet in calcium and when chlorine ranges from 2 to 10 mg/L. The 2 mg/L of chlorine, which approaches that in the drinking water of many communities, was found to increase the total serum cholesterol level about 4 mg/100 mL.[19]

This evidence suggests that consumption of chlorinated water is a possible risk factor in cardiovascular disease. Both epidemiological studies and intervention trials have revealed about a 2% increase in cardiovascular risk for each 1% increase in total cholesterol.[20,21] Thus, Wones et al.[19] concluded that "for the population at risk, even small increases in total cholesterol, if real, could have important public health implications."

Revis et al.[22] have presented data suggesting that the alternative disinfectants ClO_2 and chloramine also effect an increase in serum cholesterol levels of pigeons and rabbits. Clinical investigations funded by EPA are currently underway to assess the relevance of these preliminary findings for humans.

Cancer

The original public health concern was associated with cancer risks based on studies that showed chloroform caused liver cancer in $B6C3F_1$ mice and rats.[23] The upper-bound cancer risk (i.e., a worst-case type of analysis) indicated that the lifetime risk could reach 1 cancer per 2500.[24] In a society where "acceptable" risks from environmental exposures are often discussed in terms of 1 in 100,000 to 1 in 1 million, a risk of 1 in 2500 would be considered extremely high.

A number of epidemiological studies have supported the animal model studies indicating enhanced cancer risk from exposure to chloroform. In general, these findings have shown an association between consumption of chlorinated water and the risk of bladder cancer.[25,26] A recently completed National Cancer Institute case-control study for bladder cancer has suggested that lifetime consumption of chlorinated water increases the risk of bladder cancer by a factor of about 2 and could explain about 25–30% of the occurrence of bladder cancer in adults residing in chlorinated communities.[27]

The recognition that chlorination of water high in organic content (i.e., usually surface water) results in the formation of total THMs at levels that exceeded the U.S. EPA standard of 100 ppb has led many water utilities to seek out alternative disinfectants such as chloramine and chlorine dioxide. Use of these alternative disinfectants results in the formation of total THMs at significantly lower levels while still achieving disinfectant standards.But, compliance with EPA regulations for total THMs does not mean that the risk disappears. For example, communities with annual THM values of 80–95 ppb do not have an appreciably lower calculated cancer risk than those communities in the 100–120 ppb range even though these latter communities would be in violation of the federal standard. Nevertheless, *if the risk is real,* 1 cancer in 2500 persons exposed over their lifetimes is high. Utilities must, by law, come into compliance with EPA regulations, so they face the alternative of adopting more expensive granular activated carbon (GAC) technology to remove THM precursors or seek EPA approval for the relatively inexpensive options such as chloramine or alterations in the treatment train.

Switching from the use of a seemingly tried and true technology to alternative disinfectant technologies nevertheless presents significant challenges. Is the risk to the public health from high chloroform levels serious enough to cause hundreds of utilities affecting millions of people to switch to a different technology?

Professional judgments must be based on the reliability of the risk assessment methodologies used. Data exist from a variety of animal model studies and epidemiological evaluations. Generally, with respect to animal models, it is believed that lifetime animal studies can provide a firm foundation for making a professional judgment about whether the agent is a probable human carcinogen. Since all known human carcinogens are also animal carcinogens, it is prudent to accept, at least on a *qualitative* basis, that the agent is a probable human carcinogen.

Chloroform causes liver cancer in the mouse model only when dissolved in corn oil. If chloroform is administered in water, then the risk of liver cancer disappears.[28] The risk of kidney cancer in the rat remains whether the chloroform is dissolved in water or corn oil. So the question emerges: Which method should be used to predict risk? The general use of $B6C3F_1$ mice to make *quantitative* estimations of human risk is also of concern, since it is believed that this strain of animal markedly overstates the human risk to liver cancer. In addition, the use of biostatistical models employed by EPA for risk estimations are widely believed to err far on the side of safety due to built-in conservative assumptions. Thus, the estimated risk of cancer in the general public from consumption of 100 ppb chloroform of 1:2500 should not be taken literally. The actual risk calculated using EPA's biostatistical model could range from 1:2500 to 1:infinity, or approach zero.

Decisionmakers need to have sufficient predictive systems to markedly narrow the uncertainty range. To narrow this uncertainty and to validate previous estimates of human risk based on animal model studies, epidemiological studies have been conducted. These studies tend to compensate for many of the limitations of animal model research (i.e., the need to extrapolate from *very* high doses to realistic levels of exposure, the extrapolation from one species to another, etc.) by allowing researchers to assess risk at normal exposure levels in a relatively large number of people. Unfortunately, the epidemiological perspective has its own limitations, such as the presence of numerous and possibly important confounding variables, competing causes of death, relatively low risk at low exposures, and so on. Because of these limitations, results of epidemiological investigations often provide evidence of association only, not causality.

In terms of chlorination, the epidemiological data have reasonably confirmed the existence of up to a twofold increase risk for bladder cancer associated with consumption of chlorinated water,[27] which may result in about 1000–3000 excess bladder cancers per year in the United States. Neither of the animal models employed suggested an increased risk of bladder cancer, but different species may respond positively for tumors at different sites. In terms of our bladder cancer concern, *consumption of chlorinated water* nearly satisfied causality criteria, i.e., internal and external consistency, dose-response gradients, coherence, high relative risks, and biological plausibility. With the exception of *high relative risks,* the epidemiological data support the causal linkage.[26,27] The epidemiological approach has therefore proven very useful in assessing the relative and attributable bladder cancer risk caused by consumption of chlorinated surface water.

Trichloroacetic acid (TCA) and dichloroacetic acid (DCA), both significant by-products of the chlorination process, have been found to be carcinogenic in the B6C3F_1 mouse.[29] The levels of these agents, as noted before, closely approximate that of chloroform. The epidemiological studies assessing risk have theoretically taken into account the co-presence of chloroform, TCA, DCA, and other carcinogens possibly present. Furthermore, the levels of TCA and DCA found in some community drinking waters approach and/or slightly exceed the recent NAS[11] SNARLs (suggested no adverse response levels) of 0.50 μg TCA/L and 120 μg DCA/L for a 10-kg child.[30,31]

The evidence, though still very incomplete, is beginning to paint a picture that chronic consumption of chlorinated water will enhance the risk of developing (at least) bladder cancer by up to twofold and to enhance risk of cardiovascular disease in groups with a dietary predisposition (e.g., low calcium intake). The bladder cancer risk seems generally limited to those residing for more than 40 years in communities where drinking water was chlorinated. For individuals whose residence is less than 40 years, there appears to be no obvious enhanced risk. The capacity for epidemiological studies to quantify the contribution of chlorinated drinking water to morbidity and mortality related to the cardiovascular system will be difficult, given the modest increase in serum cholesterol in a high-risk subgroup of the population and the variety of confounding variables.

Current findings indicate that consumption of chlorinated water may nearly double bladder cancer risk in those consuming $\geq$1.5 L/day for $\geq$40 years and have a negative impact on the cardiovascular system. Given that bladder cancer has a prolonged latency and a modest relative risk and that a blood cholesterol increase of 4 mg due to chlorine in high-risk groups could be offset by prudent dietary practices, there is no immediate public health emergency to demand a switch from chlorine disinfection to any of the known alternatives. Nevertheless, the public health community should be concerned with the findings of enhanced bladder cancer and cardiovascular risks associated with chlorinated water.

Treatment Train Options

It is clear that the issue is not whether to disinfect drinking water but how to disinfect in such a way to ensure the protection of the public from infectious disease while minimizing or preventing the occurrence of chronic health effects due to the disinfectant itself or its by-products in a cost-effective way. Since chlorination of surface waters high in organic content clearly has its advantages and disadvantages, what are the practical options that communities face?

In addition to disinfection, its most important activity, chlorination has a number of functions in water treatment. Aqueous chlorine serves as an oxidant and coagulant, and it also helps to control taste and odors in the treated water. Chlorine also reduces microbial growths in transmission mains, treatment tanks, and filters. Finally, it maintains the disinfection characteristic of the treated water in the distribution system.[32] For chlorine to fulfill these multiple functions, chlorination is usually needed at two or three points of application, including a prechlorination dose, another dose to enhance disinfection directly after filtration, and a final adjusting dose as the water enters the distribution system.

Many utilities have tried to make modifications in various aspects of the treatment train to minimize haloform formation while retaining many benefits of chlorination. The modification that has been of greatest significance has been to move the point of strong initial chlorination from the entrance of the water treatment plant to a point following the coagulation and flocculation stages and even sometimes after sedimentation.

This delaying of the chlorine treatment modification has resulted in significantly lower chlorine demand and total THM formation potential (about 50–60%),[32] but results in the loss of some of the benefits of chlorination. When chlorine is used later in the treatment process, its function of enhancing coagulation via oxidation is lost. Likewise, its role in preventing biological growths in the flocculation basins is also lost. To compensate, it may be necessary to employ larger doses of coagulants such as alum, and the flocculation basins may have to be cleaned out more frequently. Other disadvantages include the diminished capacity of later chlorine treatment to reduce odorants and the greater challenge of achieving disinfection because of a diminished reaction time due to later application.[32]

Another modification to the typical chlorination process in order to reduce the production of THMs is the application of ammonia to convert residual free aqueous chlorine to chloramine as the treatment water enters into the distribution system. This adaptation has resulted in reduction of haloforms from 25% to 50%, depending on the time of transit through the distribution system and in the extent of the previous chlorine residual.

By combining the above two modifications, i.e., moving the point of chlorination and adding ammonia to form chloramine, some water utilities have been able to reduce the total THM values by 60–70%. Such a reduction has frequently brought communities' THM levels in drinking water below the 100 μg/L national primary drinking water standard of EPA.

A practical disadvantage of the use of chloramine for residual disinfection is that it is less effective than aqueous chlorine for prevention of microorganism growth in the distribution system. In addition, chloramine is much less effective than chlorine in reducing disagreeable tastes and odors.[32] Furthermore, sufficient clinical research has unequivocally found that chloramine disinfection of water used by dialysis patients can cause hemolytic anemia.[33-36] While this observation has not led to the curtailment of the use of chloramination, it has required communities adopting chloramination to fully inform the proper medical authorities about their use of this disinfectant so that proper medical adjustments can be made.

The effective control of THM formation by use of alternative disinfectants such as chloramine and chlorine dioxide will also result in the significant side benefit of reducing other by-products such as TCA, DCA, and haloacetonitriles.[37] However, both alternative disinfectants have unresolved toxicological concerns, i.e., organochloramines as possible mutagens and/or carcinogens and chlorite/chlorate as possible hemolytic agents in human high risk groups. Of particular concern is that the residual levels of both chloramine and chlorite/chlorate are likely to significantly exceed levels recommended by the NAS Safe Drinking Water Committee[11] for noncancer endpoints. While this should create a strong desire on behalf of EPA and the public health community to better assess the health effects of these agents, millions of U.S. citizens consume community drinking water treated with chloramine and dozens of communities have switched to chlorine dioxide disinfection over the past decade yet do not report adverse health effects in the subgroups identified by the NAS as high-risk.**

The use of increased amounts of coagulant aids such as alum and cationic polyelectrolytes to enhance the removal of THM precursors is not believed to present a general health concern[38] and is justified, given coagulant aid concentrations far below NOAELs (no observed adverse

**We recognize full well that the absence of a reported health effect linked to chloramine and chlorine dioxide does not mean that such effects may not be occurring. The apparent absence of a disease that society does not look for is no basis for concluding there is no problem. For example, when states have switched from a passive to active surveillance system for giardiasis, the annual reported outbreaks increased by 350%.[10] It also suggests that EPA needs to investigate this issue more thoroughly.

effect levels). However, as in the case of chloramine, persons on dialysis treatment are at risk for developing what appears to be an aluminum-induced encephalopathy when elevated levels of aluminum are present in the drinking water.[38] Based on positive aluminum balance studies, it would appear that aluminum levels in dialysate water should not exceed 10–20 μg/L. Yet, levels of 100-fold greater than this concentration may exist in some community drinking water supplies. As with chloramine, the medical community should be informed of the occurrence of elevated levels of aluminum in the public water supply so that proper clinical adjustments can be made.

Another component of a treatment train strategy for preventing THM and other by-products of chlorination is the use of activated carbon for adsorption. Activated carbon has been generally employed as an effective adsorbent for use in water treatment with partial application for the removal of unpleasant taste and odor. However, it is now being used for the removal of a broad spectrum of organic contaminants, including pesticides and synthetic organic carbons (SOCs) as well as naturally occurring precursors of chlorinated compounds generated in water treatment itself.

The removal of THM-formation potential, i.e., total organic carbon (TOC), and total adsorbable organic chlorine (TOCl) via activated carbon, may provide a valuable way of achieving compliance with the EPA maximum contaminant level (MCL) for THMs. Trihalomethanes are poorly adsorbed on granulated activated carbon. However, if GAC were used for THM precursor removal, the service life would be far shorter than if other SOCs were chosen. THM precursors (i.e., TOC) may be found in concentrations ranging from 1 to about 10 mg/L, while SOCs are often in the range of 1–100 μg/L. This high concentration of THM precursors coupled with their low adsorbability will cause the sorptive capacity of GAC beds to be exhausted relatively quickly. In marked contrast, the strong adsorption of most SOCs and their low concentrations will cause the sorptive capacity to be exhausted much more slowly.[39]

Thus, in terms of THM by-product control, adsorption processes may play a role, depending on the specific requirements of the system. Despite the strong potential public health benefits, the use of adsorption technology presents some unresolved toxicological/microbial concerns (i.e., serving as a reservoir of bacterial multiplication or the site of the formation of novel agents such as PCBs).[40] These possible negative consequences of adsorption technology are not well characterized and insufficient data exist to comprehensively evaluate their public health implications.

pH CONTROL/SOFTENING

The use of water softeners is widespread in the United States. Water softening via soft cycle ion exchange can add up to several hundred mg/L of sodium to the drinking water.[38] Sodium may also be added to the water in one of several forms, such as NaOH, for the practice of pH control as well. The concentrations of sodium added to the water for this purpose may range from less than 10 to 100–200 mg/L.[41] The issue that emerges is: What are the likely public health effects of sodium at levels of up to several hundred milligrams per liter versus the benefits of softening and pH control?

A reasonably robust epidemiological database exists concerning the relation of sodium in drinking water and blood pressure,but at this time there is no compelling evidence that sodium levels up to several hundred milligrams per liter have a significant impact on the blood pressure of healthy school-age children. With only three exceptions, the studies were ecologic in nature and were fundamentally designed to show association rather than cause and effect. Several initial studies in the United States[42,43] noted a relation between elevated levels of sodium in drinking water and blood pressure in high school and elementary school students. Their findings were supported by Hofman et al.[44] in studies of children in the Netherlands. Calabrese and Tuthill[45] also conducted an intervention study that indicated lowering of sodium in the drinking water from 120 to 10 ppm significantly reduced blood pressure in nine-year-old girls but not similarly aged boys. These initial positive studies have generally not been confirmed in studies both in the United States and elsewhere, even when sodium levels in the drinking water approached 250–400 mg/L.[46] Furthermore, follow-up studies by Tuthill and Calabrese[43] using intervention methodologies did not find any effect of sodium on the blood pressure of normal children. In one case, they added 800 mg of sodium per day to the diet for nine weeks and found no increase in blood pressure,[47] while a later study in which a community's sodium level in water decreased from 120 to 35 ppm had no impact on blood pressure nine months after the switch over to the lower sodium water.[46]

These studies involved only healthy children, and the ingestion study in which sodium was added to the diet excluded possible high-risk children. Thus, the database that exists, while extremely useful, does not allow regulatory officials to make confident judgments about the responses of individuals who are salt-sensitive and are restricted to low (1000 mg/day) or very low (500 mg/day) sodium diets. Nor does the present information address whether black males, who are acknowledged to be a subgroup of the population at enhanced risk to developing hypertension, are affected by elevated levels of sodium in drinking water.

When using sodium hydroxide for pH control, the principal concern is that of the leaching of lead from piping due to acidic water versus an excess addition of sodium to reduce the acidity of water. With lead, the concern is largely that of neurologic impairment in children; with sodium, an adverse impact on blood pressure, especially in salt-sensitive individuals. It would be ideal if one did not have to choose between neurologic damage and increased blood pressure. In fact, if pH control were carried out using calcium hydroxide, no additional sodium would be added. Furthermore, the extra calcium provided might be beneficial nutritionally. In some cases, using calcium hydroxide is both technologically and economically feasible.[46]

In choosing between the lead and excessive sodium, the elimination of lead is the primary objective, principally because lead irreversibly affects the central nervous system (CNS) of the very young. While all functions are important, those involving capacities of the CNS are most critical. The effects are likely to occur immediately after exposure and continue throughout life. Thus, water treatment practices designed to minimize exposure to lead should be sustained. If certain softening practices require that the community be exposed to 100–200 mg/L of excess sodium, it could be justified (with proper public notification) in light of the significance of lead as a neurotoxic agent, along with the ambivalent scientific literature associated with moderately elevated levels of sodium in drinking water and blood pressure.

The control of pH to prevent the release of excessive copper is of minor concern, since copper exposure at several mg/L is not believed to present a significant public health risk. While the U.S. EPA has cited a possible concern for G-6-PD deficient*** individuals,[48] there are no epidemiological data to validate their concern. Presumably, pH control for lead will also markedly diminish elevated levels of copper, therefore making copper exposure a nonsignificant issue.

When adding sodium to soften water, the principal issue is markedly different: not health versus health, but health versus convenience and economic factors. As anyone who has lived in a high hard water area knows, it is difficult to wash clothes, hair, and so on. The use of softeners has been of tremendous practical value. The issue then is not to do away with softening devices but to design their use for all washing purposes, leaving the drinking water unsoftened. Some might argue that the epidemiologic literature is sufficiently unconvincing that low sodium levels in drinking water is a health hazard and that to recommend an

***G-6-PD deficiency is a sex-linked heritable condition found in 13% of American black males and in males of Mediterranean descent. Persons with this condition may be at increased risk of developing hemolytic anemia following exposure to a number of medicinal and industrial oxidizing agents.

unsoftening component design would be a very costly approach with insufficient justification. While this appears to be the type of issue that reasonable people may differ on, the action of *not* softening the drinking water is a reasonable one and consistent with the literature indicating that Americans receive far greater amounts of sodium than their physiological requirement. There are relatively large numbers of salt-sensitive individuals who may develop cardiovascular disease later in life and do not need additional potential physiological stress. Any population-based increase in blood pressure can have an important adverse public health impact.[49]

It is well known that cement piping may contribute asbestos fibers to drinking water, especially where the water supply is aggressive. In Winnipeg, which has aggressive water, the distribution has been reported to have 6.5×10^6 fibers/L.[50] It is estimated that there are 20,000 miles of asbestos cement pipe in the United States.[51] While the epidemiological evidence is limited in this area, the fact that more than a billion feet of asbestos cement pipe could have been laid in this country to serve community drinking water illustrates the historical lack of planning, testing, and oversight for this potential health problem. While the jury remains to be heard on asbestos in drinking water, this is not the position we as a country should be in.

Fortunately, asbestos fibers can be removed with a high degree of efficiency via the use of various filters. Removal efficiencies of greater than 99% have been reported.[52] However, removal of 99% of the fibers in Winnipeg would still yield 6.5×10^4 fibers/L.

DISCUSSION

Dealing with Change

Important changes have been made in the treatment of drinking water over the past decade, principally as a response to the EPA regulation of THMs. In an effort to reduce THMs, considerable modifications in the basic treatment train have been made. The challenge is that dozens of communities are implementing various alternative approaches for disinfection—most notably, chloramine, followed by chlorine dioxide, with ozone a very distant third—and the research community is playing catch-up in this ongoing, uncontrolled societal experiment affecting millions of people. This flurry of changing water treatment techniques and research to detect associated effects on human health is principally in response to the federal primary drinking water standards for THMs.

The proliferation of research findings over the past decade into the understanding of the chlorination-THM phenomena has led to striking insights for potentially significant improvement of drinking water quality in the United States, have also revealed that the problem of chlorination-induced THM formation was only the tip of the chlorine by-product iceberg. In fact, the public health concerns are significantly greater now. In 1976, the scientific-public health community was concerned principally with chloroform and its closely related chemical cousins. Now it is known there is a bewildering array of other chlorination by-products formed along with chloroform. Preliminary toxicological returns on some of these by-products, such as TCA and DCA, reveal them to have carcinogenic properties. Most of the agents have not been adequately assessed in a toxicological sense. Other research has revealed that chloramine, chlorine dioxide, and ozone, which reduce significantly the formation of THMs, also produce an array of disinfectant by-products, the toxicological implications of which remain to be determined.

On the positive side, a number of important technical advances have been spawned. The decentralized and somewhat autonomous role of individual communities and the highly diverse nature of communities and their water supplies have led to a diverse series of intriguing engineering approaches. For example, by early 1985, 24 U.S. water treatment plants were using ozone for a variety of purposes, including THM control, disinfection, flocculation, and taste and odor control, while in 1977 only two plants were using ozone for only one purpose, taste and odor control.[53]

Given the impressive expansion of the scientific database in the area of chemical disinfectants, by-product chemistry, analytical methods and treatment processes, and toxicological assessment, it appears quite clear that the EPA opened a scientific Pandora's box with the discovery and regulation of THMs. However, the scientific understanding of the problem is under increasing intellectual control, since it is much more definable and is better understood, and practical solutions are being tested.

Troubling Dimensions/Missed Opportunities

Once its criteria (e.g., disinfection) were satisfied, the EPA has been able to permit water utilities to adopt drinking water disinfectant treatments that result in the exposure of millions of people to agents at micrograms to milligrams per liter per day without requiring systematic toxicological evaluations of the agents or their by-products. Now there is considerable interest and concern about the use of the various alternative disinfectants and the quality of the database.

This lack of planning, testing, and evaluation prior to approval for a community to use a disinfectant for the drinking water is not in the best interests of society. That EPA has repeatedly approved the use of alternative disinfectants while having an insufficient database for estimating human risks runs counter to the types of conceptual requirements that EPA makes for pesticide manufacturers and companies affected by the Toxic Substances Control Act (TSCA). Furthermore, there are numerous circumstances in which communities adopt unique alternatives in their treatment processes. These circumstances often present unusual opportunities for epidemiological intervention studies. Yet there appears to be little effective communication concerning these natural experiments between regulatory offices and the Office of Research and Development. Consequently, epidemiologic studies of potentially great value for use in regulatory affairs are frequently not considered, and great opportunities are routinely missed.

Cost Considerations: Large and Small System Differences

Economic analyses have played and will continue to play an influential, if not dominating, role in the final decision by a community with respect to the selection of their drinking water treatment process. Proponents of various technologies such as ozonation, chlorine dioxide, GAC adsorption, and other treatments have often presented cost comparisons. However, each system is unique and requires an analysis of its water quality and types of possible treatment options and costing components.

> For example, the City of Los Angeles conducted extensive cost analyses of alternative treatment processes for controlling turbidity and taste and odor, reducing THMs, and assuring bacterial and viral disinfection.[54] Their final choice was between ozonation and chlorination. By integrating ozonation into the filter plant design, the ozone option resulted in a lower capital cost. Annual operation and maintenance costs for ozone were greater than for chlorine, but the plant construction costs using chlorine exceeded 50 years of the increased ozone operation and maintenance costs. The cost of chlorine dioxide for this plant was estimated to be twice that of ozone.[54] Thus, Los Angeles decided on the ozone option.

This example is not intended to be an endorsement for ozonation, but to show the degree to which cost factors are real concerns and play a significant role.

Other factors for selecting one option over another include compliance with turbidity, taste and odor, THM levels, and disinfection. The emerging toxicology data is beginning to suggest that disinfectant by-products, the level of residual chlorine in the drinking water, and other factors should also influence the decisionmaking processes of a community.

The solutions of large communities are not easily transferable to small systems. Small systems often are not able to take advantage of economies of scale in treatment techniques, resulting in much greater per capita costs. For example, initial estimations of the preliminary costs for controlling VOCs in public water supplies indicate that the monthly per-household cost increases for such treatment will be two- to sixfold higher in small systems than larger systems independent of the technology selected (Table 4). In addition, small systems frequently have less latitude in seeking alternative water supplies and thus are more likely than large systems to experience local water quality problems. Some treatment techniques are not financially available to small systems with limited resources. The problems of small systems are acknowledged by EPA with respect to THMs. In this instance, the EPA solution was to exempt all systems serving less than 10,000 from the THM standard.[55]

In 1980, nearly 14,000 community water systems were in violation of one or more Interim Primary Drinking Water Regulations. Two years later, in 1982, over 70,000 violations of the interim regulations were recorded by 20,000 community water systems. While most violations were of monitoring and reporting standards, EPA estimated that more than 9000 systems needed to improve treatment facilities in order to meet health-related drinking water standards.[56]

EPA has noted that compliance with drinking water regulations is a problem mostly for small systems (serving less than 3300 people). EPA

Table 4. Summary of Increased Per-Household Costs for Reducing Trichloroethylene (a VOC) Levels in Drinking Water

	Population Served		
Technology	**100–499**	**1,000–2,499**	**10,000–24,999**
	90% Removal (500 mg/L to 50 mg/L)		
Aeration (Packed Tower)	$4.65–6.49	$2.04–2.58	$0.72–0.92
Aeration (Diffused Air)	9.76	5.04	2.41
Adsorption (GAC)	13.48	7.62	2.03
	99% Removal (500 mg/L to 5 mg/L)		
Aeration (Packed Tower)	$5.11–7.43	$2.34–12.00	$0.85–1.13
Aeration (Diffused Air)	13.48	8.23	3.81
Adsorption (GAC)	13.95	7.87	6.32

Source: Harker.[55]

reports that in 1982 the microbiological requirements were not met by 10% of the smaller systems.[57] Between 1500–3000 systems exceeded MCLs for certain inorganic contaminants,[57] with the problems centering on arsenic, barium, lead (from pipe or solder corrosion), fluoride, and nitrate. According to Harker,[55] the inability of small community systems to pay for the needed improvements in order to achieve compliance with the primary drinking water regulations is an important cause of the high noncompliance rates. In fact, EPA indicates that of the communities violating one or more MCL, 63% (8700), would need "relatively expensive treatment improvement" to avoid violations with the primary standards. About 3300 of the noncompliance systems are believed to be "unable to make necessary improvements using conventional local financing." Most of these systems (91%) serve less than 500 consumers.[58]

EPA faces a fundamental challenge: how to assist public water systems in their efforts to provide water of acceptable quality to their consumers. Given that the principles of economies of scale will pose nearly insurmountable barriers to many small systems to compliance with present and proposed EPA primary drinking water standards, what options are available to them? First, just because a community drinking water system is in violation of an EPA standard does not automatically mean there is a genuine public health risk. In many instances there is a generous margin of safety built into the MCLs. Thus, if an MCL were violated, an adverse health effect may not necessarily be expected even in high-risk groups. A number of factors are important, including the actual concentration of impurities in the water, the built-in margin of safety, duration of exposure, the nature of the population, and others. The mere recording of a violation should not be viewed as a condemnation of the water source and an absolute requirement to adopt very expensive treatment technology. Each water treatment system needs to be evaluated on its own merits within the overall context of the goals of the Safe Drinking Water Act.

If a review finds that, indeed, the violation poses a significant risk to the public health, then the community should be assisted by the state and EPA in the technical evaluation of the drinking water system. In concert with the technical evaluation, plans need to be developed to assist the community financially in the upgrading of the system for compliance with federal requirements. Harker[55] has proposed that bottled water could serve as a permanent supply of potable water for an entire small community or noncommunity system where the cost of upgrading the system cannot be met: "The use of bottled water where the financial capacity of the system is insufficient to support the upgrading of centralized treatment would accomplish the purposes of the Safe Drinking

Water Act considerably better than merely exempting small systems from various contaminant regulations...." Furthermore, the act actually permits the EPA to authorize the use of bottled water, where appropriate, to achieve the goals of the act.

The weighing of health risks and compliance costs is a difficult task and a task made more difficult for the small systems facing serious water quality problems. Consequently, the agency needs to develop flexible approaches for achieving the goals of the Safe Drinking Water Act, especially in the case of small systems.

Developing a Perspective

The state of drinking water quality is highly variable in the United States. As noted before, it is a function of multiple variables, including the source of water, available alternative sources, the type of distribution system, the size of the population, and economic and technical resources available. Drinking water problems and solutions or partial solutions are seen within a risk management framework, especially at the local level. The final product must be one that will reflect an acceptable balance to the community of health, aesthetics, and economic components.

That EPA establishes MCLs for numerous agents is a compromise between the ideal and reality. For example, the establishment of an MCL for total THMs of 100 ppb, which has been estimated to carry an upper-bound cancer risk of 1:2500 over a lifetime exposure, is clearly not the ideal situation even if the real risk proves to be much lower than the above estimate. In the broader sense, all utilities with a high-THM problem should be striving to reduce it to as low a level as possible, but in effect, what happens is that utilities will work hard to lower it to satisfy an EPA *regulation*—not to lower it as much as possible. Thus, utilities will not get excited about elevated levels of TCA, DCA, haloacetonitriles, or other chlorination by-products—even though they are animal carcinogens—unless EPA makes it an enforceable regulation.

It is technologically feasible to markedly lower the level of toxic or carcinogenic agents in most community drinking water sources. However, the cost of thoroughly studying a system, designing solutions, and installation and operation of the necessary treatments or processes are normally beyond the reach of consumers' ability to pay. Most systems try to compromise between doing little or nothing and applying techniques they cannot afford. They try to modify slightly the system they have and make improvements sufficient to satisfy EPA. Thus, one finds a variety of relatively straightforward treatment train modifications of amounts and types of oxidants and coagulant aids, and the use of alternative disinfectants such as chloramine. These approaches are those of systems trying to provide drinking water with sufficient disinfectant to prevent risk of infectious diseases and minimize exposure to chemical toxins while not greatly increasing the monthly water bill. In effect, the water utility companies have achieved these goals—and should be commended for it. However, this should not delude us into believing that the drinking water quality in the United States as a whole is without significant public health impact. In general, U.S. water quality and treatment practices are sufficient to get the job done—reasonably well. On a 10-point scale it should probably receive a 7.0 or 8.0. It scores its greatest points for ensuring a high disinfectant quality to the water. But it has serious deficiencies that affect the long-term health of the population. Given other important needs, how much more should be allocated to cleaner and safer drinking water?

To achieve the goal of the near elimination of the spread of bacterial and viral diseases via drinking water in this country, U.S. residents are also believed to be paying a health price as well. Disinfectant processes, especially chlorination, may create new and potentially toxic and carcinogenic products in the water that users consume. Through careful study and testing, EPA and the private sector have learned to significantly reduce, but not eliminate, some of the products of these disinfectants. Further study is likely to provide a better understanding of the actual risks to the public caused by these agents and improved engineering solutions to effect a more healthful product.

Society needs to hold on to its achievement of the near elimination of waterborne microbiological disease while at the same time acquiring a better understanding of how to minimize the long-term toxicity concerns. In the case of drinking water treatment technology, it is essential that society not throw the baby out with the bath water. On the other hand, the consumer should not be misled into believing that he or she has the best product money can buy. Individuals may buy certain types of bottled water that is believed to be better than what they get out of the tap but in reality is completely unregulated. It is possible that privately sold bottled water or various and correctly operated point-of-use devices could provide a significant improvement on what is actually being consumed. However, to make that choice wisely and to ensure that it is implemented properly requires extra cost, extra study, and good judgment.

The federal government and the private water utilities must continue to recognize and accept the challenge to improve the quality of water supplied to consumers. The present society is the beneficiary of past and present successes, but has begun to recognize that there are problems inherent in current technological approaches. A widespread risk communication program for the public about the quality of the water, its methodologies, and treatment problems is essential for enhanced community understanding and the development of realistic solutions to improve American drinking water quality.

REFERENCES

1. Manwaring, J. F. "Public Drinking Water and Chemicals," in *Safe Drinking Water: The Impact of Chemicals on a Limited Resource,* R. G. Rice, Ed. (Chelsea, MI: Lewis Publishers, Inc., 1985), pp. 21–31.
2. Hutton, J. T. "Bottled Water: An Alternative Source of Safe Drinking Water," in *Safe Drinking Water: The Impact of Chemicals on a Limited Resource,* R. G. Rice, Ed. (Chelsea, MI: Lewis Publishers, Inc., 1985), pp. 33–42.
3. Galbraith, P. "National Survey of Environmental Health Priorities of State Public Health Commissioners," Connecticut Department of Health Services, Hartford, CT (1987).
4. White, G. D. *Handbook of Chlorination* (New York: Van Nostrand Reinhold Company, 1972).
5. Akin, E. W., and J. C. Hoff. "Microbiological Risks Associated with Changes in Drinking Water Disinfection Practices," *Water Chlorination: Chemistry, Environmental Impact and Health Effects, Vol. 5,* R. L. Jolley, R. J. Bull, W. P. Davis, S. Katz, M. H. Roberts, Jr., and V. A. Jacobs, Eds. (Chelsea, MI: Lewis Publishers, Inc., 1985), pp. 99–110.
6. "Manual for Evaluating Public Drinking Water Supplies," U.S. Public Health Service, USPHS Publ. 182 (1969). [Reprinted as EPA-430/9–75–011 (Washington, DC: U.S. Environmental Protection Agency, 1971, 1974, and 1975)].
7. Akin, E. W., G. Berg, N. A. Clarke, R. Culp, R. S. Englebrecht, E. H. Lennette, T. Metcalf, J. W. Mosley, H. E. Pearson, R. Sullivan, and H. W. Wolf. "Viruses in Drinking Water—Committee Report," *J. AWWA* 71(8):441–444 (1979).
8. Akin, E. W. "A Review of Infective Dose Data for Enteroviruses and other Enteric Microorganisms in Human Subjects," in *Microbial Health Consideration of Soil Disposal of Domestic Wastewater: Proceedings,* L. W. Canter, E. W. Akin, J. F. Kreissl, and J. F. McNabb, Eds. EPA-600/9–83–017 (Cincinnati, OH: U.S. Environmental Protection Agency, 1983), pp. 304–322.
9. Schiff, G. M., G. M. Stefanovic, E. C. Young, D. S. Sander, J. K. Pennekamp, and R. L. Ward. "Studies of Echovirus 12 in Volunteers; Determination of Minimum Infective Dose and Effect of Previous Infection on Infectious Dose," *J. Infect. Dis.* 150:858–866 (1984).
10. "Drinking Water Criteria Document for *Giardia,*" U.S. Environmental Protection Agency, National Technical Information Service, Springfield, VA (1984).
11. *Drinking Water and Health, Vol. 7* (Washington, DC: National Academy of Sciences, 1987).

12. Hejkal, T. W., B. Keswick, R. L. LaBelle, C. P. Gerba, Y. Sanchez, G. Dreesman, B. Hafkin, and J. L. Melnick. "Viruses in a Community Water Supply Associated with an Outbreak of Gastroenteritis and Infectious Hepatitis," *J. AWWA* 74(6):318–321 (1982).
13. Lippy, E. C., and S. C. Waltrip. "Waterborne Disease Outbreaks—1946–1980: A Thirty-Five-Year Perspective," *J. AWWA* 76(2):60–67 (1984).
14. Blabaum, C. J., and M. S. Nichols. "Effects of Highly Chlorinated Drinking Water on White Mice," *J. AWWA* 48:1503–1506 (1956).
15. Druckery, H. "Chlorinated Drinking Water Toxicity Studies in Seven Generations of Rats," *Food Cosmet. Toxicol.* 6:147–152 (1968).
16. Revis, N. W., T. R. Osborne, G. Holdsworth, and P. McCauley. "Effect of Chlorinated Drinking Water on Myocardial Structure and Functions in Pigeons and Rabbits," in *Water Chlorination: Chemistry, Environmental Impact and Health Effects, Vol. 5,* R. L. Jolley, R. J. Bull, W. P. Davis, S. Katz, M. H. Roberts, Jr., and V. A. Jacobs, Eds. (Chelsea, MI: Lewis Publishers, Inc., 1985), pp. 365–371.
17. Revis, N. W., B. H. Douglas, P. T. McCauley, H. P. Witschi, and R. J. Bull. "The Relationship of Drinking Water Chlorine to Coronary Atherosclerosis," *Pharmacologist* 25:732 (1983).
18. Dresser, C. M., M. D. Carroll, and S. Abraham. *Vital and Health Statistics,* Series II, National Health and Nutrition Examination Surveys, National Center for Statistics, Department of Health and Human Services, Hyattsville, MD (1984).
19. Wones, R. G., L. Mieczkowski, and L. A. Frohman. "Drinking Water and Human Lipid and Thyroid Metabolism," in *Water Chlorination: Chemistry, Environmental Impact and Health Effects, Vol. 6.* Forthcoming in 1989.
20. Stamler, J., D. Wentworth, and J. D. Neatong. "Is Relationship Between Serum Cholesterol and Risk of Premature Death from Coronary Heart Disease Continuous and Graded? *J. Am. Med. Assoc.* 256(20):2823–8 (1986).
21. The Lipid Research Clinic. "The Lipid Research Clinic's Coronary Primary Prevention Trial Results," *J. Am. Med. Assoc.* 251:351–374 (1984).
22. Revis, N. W., P. McCauley, R. Bull, and G. Holdsworth. "Relationship of Drinking Water Disinfectants to Plasma Cholesterol and Thyroid Hormone Levels in Experimental Studies," *Proc. Nat. Acad. Sci., U.S.* 83:1485–1489 (1986).
23. "Carcinogenesis Bioassay of Chloroform," National Cancer Institute, NTIS No. PB264018/AS (1976).
24. "National Interim Primary Drinking Water Regulations," U.S. Environmental Protection Agency, 40 *U.S. Code of Federal Regulations* 141 (June 24, 1977).
25. Crump, K. S., and H. A. Guess. "Drinking Water and Cancer: Review of Recent Epidemiological Findings and Assessment of Risks," *Ann. Rev. Public Health* 3:339–357 (1982).
26. Beresford, S. A. A. "Epidemiologic Assessment of Health Risks Associated with Organic Micropollutants in Drinking Water," in *Organic Carcinogens in*

Drinking Water: Detection, Treatment and Risk Assessment, N. Ram, E. J. Calabrese, and R. Christman, Eds. (New York: John Wiley & Sons, Inc., 1986), pp. 373–404.

27. Cantor, R. P., R. Hoover, P. Hartaige, T. Mason, and D. Silverman. "Bladder Cancer, Tap Water Consumption, and Drinking Water Source," presented to Sixth Annual Conference on Water Chlorination, Oakridge, TN, May 3–8, 1987.
28. Jorgensen, T., E. Meierhenry, C. F. Rushbrook, R. J. Bull, M. Robinson, and C. E. Whitmire. "Carcinogenicity of Chloroform in Drinking Water to Male Osborne-Mendel Rats and Female B6C3F1 Mice," *Fund. Appl. Toxicol.* 5:760–769 (1985).
29. Herren-Freund, S. L., M. A. Pereira, and G. Olson. "The Carcinogenicity of Trichloroethylene and Its Metabolites, Trichloroacetic Acid and Dichloroacetic Acid, in Mouse Liver," *Toxicol. Appl. Pharmacol.* 90(2):183–189 (1987).
30. Miller, J. W., and P. C. Uden. "Characterization of Nonvolatile Aqueous Chlorination Products of Humic Substances," *Environ. Sci. Technol.* 17:150–157 (1983).
31. Norwood, D. L., G. P. Thompson, J. D. Johnson, and R. F. Christman. "Monitoring Trichloroacetic Acid in Municipal Drinking Water," in *Water Chlorination: Chemistry, Environmental Impact and Health Effects, Vol. 5,* R. L. Jolley, R. J. Bull, W. P. Davis, S. Katz, M. H. Roberts, Jr., and V. A. Jacobs, Eds. (Chelsea, MI: Lewis Publishers, Inc., 1985), pp. 1115–1122.
32. Morris, J. C. "Aqueous Chlorine in the Treatment of Water Supplies," in *Organic Carcinogens in Drinking Water: Detection, Treatment and Risk Assessment,* N. Ram, E. J. Calabrese, and R. Christman, Eds. (New York: John Wiley & Sons, Inc., 1985), pp. 33–54.
33. Kjellstrand, C. M., J. W. Eaton, Y. Yawata, H. Swofford, C. F. Kolpin, T. J. Buselmeier, B. Von Hartitzsch, and H. S. Jacob. "Hemolysis in Dialyzed Patients Caused by Chloramines," *Nephron* 13:427–433 (1974).
34. Jacob, H. S., J. W. Eaton, and Y. Yawata. "Shortened Red Cell Survival in Uremic Patients: Beneficial and Deleterious Effects of Dialysis," *Kidney Int.* (Suppl.) 2:139–143 (1975).
35. Eaton, J. W., C. F. Kolpin, C. M. Kjellstrand, and H. S. Jacob. "Chlorinated Urban Water: A Cause of Dialysis-Induced Hemolytic Anemia," *Science* 181:463–464 (1973).
36. Yawata, Y., C. Kjellstrand, T. Buselmeier, R. Howe, and H. Jacob. "Hemolysis in Dialyzed Patients: Tap Water-Induced Red Blood Cell Metabolic Deficiency," *Trans. Am. Soc. Artif. Int. Organs* 18:301–304 (1972).
37. Stevens, A. A., and J. M. Symons. "Alternative Disinfection Processes," in *Organic Carcinogens in Drinking Water,* N. Ram, E. J. Calabrese, and R. Christman, Eds. (New York: John Wiley & Sons, Inc., 1985), pp. 265–290.
38. "Comparative Health Effects Assessment of Drinking Water Treatment Technologies Report to Congress," U.S. Environmental Protection Agency, Washington, DC (1988).
39. DiGiano, F. A. "Removal of Organic Contaminants in Drinking Water by

Adsorption," in *Organic Carcinogens in Drinking Water: Detection, Treatment and Risk Assessment,* N. Ram, E. J. Calabrese and R. Christman, Eds. (New York: John Wiley & Sons, Inc., 1985), pp. 237–264.

40. Voudrias, E. A., R. A. Larson, V. L. Snoeyink, and A. S.-C. Chen. "Activated Carbon: An Oxidant Producing Hydroxylated PCBs," in *Water Chlorination: Chemistry, Environmental Impact and Health Effects, Vol. 5,* R. L. Jolley, R. J. Bull, W. P. Davis, S. Katz, M. H. Roberts, Jr., and V. A. Jacobs, Eds. (Chelsea, MI: Lewis Publishers, Inc., 1985), pp. 1313–1328.
41. Calabrese, E. J., and R. W. Tuthill. "Water Treatment Processes as a Contributor to Elevated Levels of Sodium in Drinking Water, *J. Environ. Sci. Health* A13(3):253–260 (1978).
42. Calabrese, E. J., and R. W. Tuthill. "The Massachusetts Blood Pressure Studies–I," *Adv. Modern Environ. Toxicol.* 9:1–10. Concurrently published in *Toxicol. Ind. Health* 1(1):1–10 (1985a).
43. Tuthill, R. W., and E. J. Calabrese. "The Massachusetts Blood Pressure Studies–II," *Adv. Mod. Environ. Toxicol.* 9:11–18. Concurrently published in *Toxicol. Ind. Health* 1(1):11–19 (1985a).
44. Hofman, A. "Blood Pressure and Sodium Intake: Evidence from Two Dutch Studies," *Adv. Mod. Environ. Toxicol.* 9:43–48 (1985).
45. Calabrese, E. J., and R. W. Tuthill. "The Massachusetts Blood Pressure Studies–III," *Adv. Modern Environ. Toxicol.* 9:19–34. Concurrently published in *Toxicol. Ind. Health* 1(1):19–34 (1985b).
46. Calabrese, E. J., and R. W. Tuthill. "The Lowering of a Community's Drinking Water Sodium Concentration Has No Effect upon Adolescent Blood Pressures," final report to American Water Works Association Research Foundation (1987).
47. Tuthill, R. W., and E. J. Calabrese. "The Massachusetts Blood Pressure Studies–IV," *Adv. Mod. Environ. Toxicol.* 9:35–44. Concurrently published in *Toxicol. Ind. Health* 1(1):35–44 (1985b).
48. Moore, G. S., and E. J. Calabrese. "The Effects of Copper and Chlorite on Normal and G-6-PD Deficient Erythrocytes," *J. Environ. Pathol. Toxicol.* 4:271–280 (1980).
49. Wilkins, J., and E. J. Calabrese. "The Health Implications of a 3–5 mmHg Increase in Blood Pressure in a Community," *Adv. Mod. Environ. Toxicol.* 9:85–100 (1985).
50. Hickman, J. R. "Drinking Water: A Global Victual," in *Safe Drinking Water: The Impact of Chemicals on a Limited Resource,* R. G. Rice, Ed. (Chelsea, MI: Lewis Publishers, Inc., 1985), pp. 9–20.
51. Olson, H. L. "Asbestos in Potable Water Supplies," *J. AWWA* 68:215 (1974).
52. Cook, P. M., G. Glass, D. Marklund, and J. Tucker. "Evaluation of Cartridge Filters for Removal of Small Fibers from Drinking Water," *J. AWWA* (8):459 (1978).
53. Rice, R. G. "Ozone for Drinking Water Treatment—Evolution and Present Status," in *Safe Drinking Water: The Impact of Chemicals on a Limited*

Resource, R. G. Rice, Ed. (Chelsea, MI: Lewis Publishers, Inc., 1985), pp. 123–159.

54. Stolarik, G. "Ozonation–Direct Filtration of Los Angeles Drinking Water," presented at the International Ozone Association, 6th Ozone World Congress, Washington, DC, May 1983.
55. Harker, T. L. "Regulatory Flexibility and Consumer Options under the Safe Drinking Water Act," in *Safe Drinking Water: The Impact of Chemicals on a Limited Resource,* R. Rice, Ed. (Lewis Publishers, Inc., Chelsea, MI, 1985), pp. 209–221.
56. *Federal Register,* Vol. 48, p. 45505 (1983).
57. *Federal Register,* Vol. 48, p. 45504 (1983).
58. "Small System Compliance for Public Water Supply Systems," U.S. Environmental Protection Agency, Safe Drinking Water Act, *Federal Register* 45:40223 (June 13, 1980).

APPENDIX 1

The Safe Drinking Water Act

As Amended by the
Safe Drinking Water Act Amendments of 1986
(Enacted June 19, 1986)

Source: Adapted from a publication prepared by Camp Dresser & McKee Inc., Boston, Massachusetts. Additions have been inserted in boldface type. The repealed sections of the Act have been deleted from this version.

CONTENTS

Part C—Protection of Underground Sources of Drinking Water

Part D—Emergency Powers

Part E—General Provisions

TITLE XIV—SAFETY OF PUBLIC WATER SYSTEMS

Part A—Definitions

DEFINITIONS

Section 1401. For purposes of this title:

(1) The term "primary drinking water regulation" means a regulation which—

(A) applies to public water systems;

(B) specifies contaminants which, in the judgment of the Administrator, may have any adverse effect on the health of persons;

(C) specifies for each such contaminant either—

(i) a maximum contaminant level, if, in the judgment of the Administrator, it is economically or technologically feasible to ascertain the level of such contaminant in water in public water systems, or

(ii) if, in the judgment of the Administrator, it is not economically or technologically feasible to so ascertain the level of such contaminant, each treatment technique known to the Administrator which leads to a reduction in the level of such contaminant sufficient to satisfy the requirements of section 1412; and

(D) contains criteria and procedures to assure a supply of drinking water which dependably complies with such maximum contaminant levels; including quality control and testing procedures to insure compliance with such levels and to insure proper operation and maintenance of the system, and requirements as to (i) the minimum quality of water which may be taken into the system and (ii) siting for new facilities for public water systems.

(2) The term "secondary drinking water regulation" means a regulation which applies to public water systems and which specifies the maximum contaminant levels which, in the judgment of the Administrator, are requisite to protect the public welfare. Such regulations may apply to any contaminant in drinking water (A) which may adversely affect the odor or appearance of such water and consequently may cause a substantial number of the persons served by the public water system providing such water to discontinue its use, or (B) which may otherwise adversely affect the public welfare. Such regulations may vary according to geographic and other circumstances.

(3) The term "maximum contaminant level" means the maximum permissible level of a contaminant in water which is delivered to any user of a public water system.

(4) The term "public water system" means a system for the provision to the public of piped water for human consumption, if such system has at least fifteen service connections or regularly serves at least twenty-five individuals. Such term includes (A) any collection, treatment, storage, and distribution facilities under control of the operator of such system and used primarily in connection with such system, and (B) any collection or pretreatment storage facilities not under such control which are used primary in connection with such system.

(5) The term "supplier of water" means any person who owns or operates a public water system.

(6) The term "contaminant" means any physical, chemical, biological, or radiological substance or matter in water.

(7) The term "Administrator" means the Administrator of the Environmental Protection Agency.

(8) The term "Agency" means the Environmental Protection Agency.

(9) The term "Council" means the National Drinking Water Advisory Council established under section 1446.

(10) The term "municipality" means a city, town, or other public body created by or pursuant to State law, or an **Indian Tribe** authorized by law.

(11) The term "Federal agency" means any department, agency, or instrumentality of the United States.

(12) The term "person" means an individual, corporation, company, association, partnership, State, municipality, or Federal agency (and includes officers, employees, and agents of any corporation, company, association, State, municipality, or Federal agency).

(13) The term "State" includes, in addition to the several States, only the District of Columbia, Guam, the Commonwealth of Puerto Rico, the Northern Mariana Islands, the Virgin Islands, American Samoa, and the Trust Territory of the Pacific Islands.

(14) The term "Indian Tribe" means any Indian tribe having a Federally recognized governing body carrying out substantial governmental duties and powers over any area.

Part B—Public Water Systems

COVERAGE

Section 1411. Subject to sections 1415 and 1416, national primary drinking water regulations under this part shall apply to each public water system in each State; except that such regulations shall not apply to a public water system—

(1) which consists only of distribution and storage facilities (and does not have any collection and treatment facilities);

(2) which obtains all of its water from, but is not owned or operated by, a public water system to which such regulations apply;

(3) which does not sell water to any person; and

(4) which is not a carrier which conveys passengers in interstate commerce.

NATIONAL DRINKING WATER REGULATIONS

Section 1412. **(a)(1) Effective on the enactment of the Safe Drinking Water Act Amendments of 1986, each national interim or revised primary drinking water regulation promulgated under this section before such enactment shall be deemed to be a national primary drinking water regulation under subsection (b). No such regulation shall be required to comply with the standards set forth in subsection (b)(4) unless such regulation is amended to establish a different maximum contaminant level after the enactment of such amendments.**

(2) After the enactment of the Safe Drinking Water Act Amendments of 1986 each recommended maximum contaminant level published before the enactment of such amendments shall be treated as a maximum contaminant level goal.

(3) Whenever a national primary drinking water regulation is proposed under paragraph (1), (2), or (3) of subsection (b) for any contaminant, the maximum contaminant level goal for such contaminant shall be proposed simultaneously. Whenever a national primary drinking water regulation is promulgated under paragraph (1), (2), or (3) of subsection (b) for any contaminant, the maximum contaminant level goal for such contaminant shall be published simultaneously.

(4) Paragraph (3) shall not apply to any recommended maximum contaminant level published before the enactment of the Safe Drinking Water Act Amendments of 1986.

(b)(1) In the case of those contaminants listed in the Advance Notice of Proposed Rulemaking published in volume 47, Federal Register, page 9352, and in volume 48, Federal Register, page 45502, the Administrator shall publish maximum contaminant level goals and promulgate national primary drinking water regulations—

(A) not later than 12 months after the enactment of the Safe Drinking Water Act Amendments of 1986 for not less than 9 of those listed contaminants;

(B) not later than 24 months after such enactment for not less than 40 of those listed contaminants; and

(C) not later than 36 months after such enactment for the remainder of such listed contaminants.

(2)(A) If the Administrator identifies a drinking water contaminant the regulation of which, in the judgment of the Administrator, is more likely to be protective of public health (taking into account the schedule for regulation under paragraph (1), the Administrator may publish a maximum contaminant level goal and promulgate a national primary drinking water regulation for such identified contaminant in lieu of regulating the contaminant referred to in such paragraph. There may be no more than 7 contaminants in paragraph (1) for which substitutions may be made. Regulation of a contaminant identified under this paragraph shall be in accordance with the schedule applicable to the contaminant for which the substitution is made.

(B) If the Administrator identifies one or more contaminants for substitution under this paragraph, the Administrator shall publish in the Federal Register not later than one year after the enactment of the Safe Drinking Water Act Amendments of 1986 a list of contaminants proposed for substitutions, the contaminants referred to in paragraph (1) for which substitutions are to be made, and the basis for the judgment that regulation of such proposed substitute contaminants is more likely to be protective of public health (taking into account the schedule for regulation under such paragraph). Following a period of 60 days for public comment, the Administrator shall publish in the Federal Register a final list of contaminants to be substituted and contaminants referred to in paragraph (1) for which substitutions are to be made, together with responses to significant comments.

(C) Any contaminant referred to in paragraph (1) for which a substitution is made, pursuant to subparagraph (A) of this paragraph, shall be included on the priority list to be published by the Administrator not later than January 1, 1988, pursuant to paragraph (3)(A).

(D) The Administrator's decision to regulate a contaminant identified pursuant to this paragraph in lieu of a contaminant referred to in paragraph (1) shall not be subject to judicial review.

(3)(A) The Administrator shall publish maximum contaminant level goals and promulgate national primary drinking water regulations for each contaminant (other than a contaminant referred to in paragraph (1) or (2) for which a national primary drinking water regulation was promulgated) which, in the judgment of the Administrator, may have any adverse effect on the health of persons and which is known or anticipated to occur in public water systems. Not later than January 1, 1988, and at 3 year intervals thereafter, the Administrator shall publish a list of contaminants which are known or anticipated to occur in public water systems and which may require regulation under this Act.

(B) For the purpose of establishing the list under subparagraph (A), the Administrator shall form an advisory working group including members from the National Toxicology Program and the Environmental Protection Agency's Offices of Drinking Water, Pesticides, Toxic Substances, Ground Water, Solid Waste and Emergency Response and any others the Administrator deems appropriate. The Administrator's consideration of priorities shall include, but not be limited to, substances referred to in section 101(14) of the Comprehensive Environmental Response, Compensation, and Liability Act of 1980, and substances registered as pesticides under the Federal Insecticide, Fungicide, and Rodenticide Act.

(C) Not later than 24 months after the listing of contaminants under subparagraph (A), the Administrator shall publish proposed maximum contaminant level goals and national primary drinking water regulations for not less than 25 contaminants from the list established under subparagraph (A).

(D) Not later than 36 months after the listing of contaminants under subparagraph (A), the Administrator shall publish a maximum contaminant goal and promulgate a national primary drinking water regulation for those contaminants for which proposed maximum contaminant level goals and proposed national primary drinking water regulations were published under subparagraph (C).

(4) Each maximum contaminant level goal established under this subsection shall be set at the level at which no known or anticipated adverse effects on the health of persons occur and which allows an adequate margin of safety. Each national primary drinking water regulation for a contaminant for which a maximum contaminant level goal is established under this subsection shall specify a maximum level for such contaminant which is as close to the maximum contaminant level goal as is feasible.

(5) For the purposes of this subsection, the term "feasible" means feasible with the use of the best technology, treatment techniques and other means which the

Administrator finds, after examination for efficacy under field conditions and not solely under laboratory conditions, are available (taking cost into consideration). For the purpose of paragraph (4), granular activated carbon is feasible for the control of synthetic organic chemicals, and any technology treatment technique, or other means found to be the best available for the control of synthetic organic chemicals must be at least as effective in controlling synthetic organic chemicals as granular activated carbon.

(6) Each national primary drinking water regulation which establishes a maximum contaminant level shall list the technology, treatment techniques, and other means which the Administrator finds to be feasible for purposes of meeting such maximum contaminant level, but a regulation under this paragraph shall not require that any specified technology, treatment technique, or other means be used for purposes of meeting such maximum contaminant level.

(7)(A) The Administrator is authorized to promulgate a national primary drinking water regulation that requires the use of a treatment technique in lieu of establishing a maximum contaminant level, if the Administrator makes a finding that it is not economically or technologically feasible to ascertain the level of the contaminant. In such case, the Administrator shall identify those treatment techniques which, in the Administrator's judgment, would prevent known or anticipated adverse effects on the health of persons to the extent feasible. Such regulations shall specify each treatment technique known to the Administrator which meets the requirements of this paragraph, but the Administrator may grant a variance from any specified treatment technique in accordance with section 1415(a)(3).

(B) Any schedule referred to in this subsection for the promulgation of a national primary drinking water regulation for any contaminant shall apply in the same manner if the regulation requires a treatment technique in lieu of establishing a maximum contaminant level.

(C)(i) Not later than 18 months after the enactment of the Safe Drinking Water Act Amendments of 1986, the Administrator shall propose and promulgate national primary drinking water regulations specifying criteria under which filtration (including coagulation and sedimentation, as appropriate) is required as a treatment technique for public water systems supplied by surface water sources. In promulgating such rules, the Administrator shall consider the quality of source waters, protection afforded by watershed management, treatment practices (such as disinfection and length of water storage) and other factors relevant to protection of health.

(ii) In lieu of the provisions of section 1415 the Administrator shall specify procedures by which the State determines which public water systems within its jurisdiction shall adopt filtration under the criteria of clause (i). The State may require the public water system to provide studies or other information to assist in this determination. The procedures shall provide notice and opportunity for public hearing on this determination. If the State determines that filtration is required, the State shall prescribe a schedule for compliance by the public water system with the filtration requirement. A schedule shall

require compliance within 18 months of a determination made under clause (iii).

(iii) Within 18 months from the time that the Administrator establishes the criteria and procedures under this subparagraph, a State with primary enforcement responsibility shall adopt any necessary regulations to implement this subparagraph. Within 12 months of adoption of such regulations the State shall make determinations regarding filtration for all the public water systems within its jurisdiction supplied by surface waters.

(iv) If a State does not have primary enforcement responsibility for public water systems, the Administrator shall have the same authority to make the determination in clause (ii) in such State as the State would have under that clause. Any filtration requirement or schedule under this subparagraph shall be treated as if it were a requirement of a national primary drinking water regulation.

(8) Not later than 36 months after the enactment of the Safe Drinking Water Act Amendments of 1986, the Administrator shall propose and promulgate national primary drinking water regulations requiring disinfection as a treatment technique for all public water systems. The Administrator shall simultaneously promulgate a rule specifying criteria that will be used by the Administrator (or delegated State authorities) to grant variances from this requirement according to the provisions of section 1451(a)(1)(B) and 1415(a)(3). In implementing section 1442(g) the Administrator or the delegated State authority shall, where appropriate, give special consideration to providing technical assistance to small public water systems in complying with the regulations promulgated under this paragraph.

(9) National primary drinking water regulations shall be amended whenever changes in technology, treatment techniques, and other means permit greater protection of the health of persons, but in any event such regulations shall be reviewed at least once every 3 years. **Such review shall include an analysis of innovations or changes in technology, treatment techniques, or other activities that have occurred over the previous 3-year period and that may provide for greater protection of the health of persons. The findings of such review shall be published in the Federal Register. If, after opportunity for public comment, the Administrator concludes that the technology, treatment techniques, or other means resulting from such innovations or changes are not feasible within the meaning of paragraph (5), an explanation of such conclusion shall be published in the Federal Register.**

(10) National primary drinking water regulations promulgated under this subsection (and amendments thereto) shall take effect eighteen months after the date of their promulgation. Regulations under subsection (a) shall be superseded by regulations under this subsection to the extent provided by the regulations under this subsection.

(11) No national primary drinking water regulation may require the addition of any substance for preventive health care purposes unrelated to contamination of drinking water.

(c) The Administrator shall publish proposed national secondary drinking water regulations within 270 days after the date of enactment of this title. Within 90 days after publication of any such regulation, he shall promulgate such regulation with such modifications as he deems appropriate. Regulations under this subsection may be amended from time to time.

(d) Regulations under this section shall be prescribed in accordance with section 553 of title 5, United States Code (relating to rulemaking), except that the Administrator shall provide opportunity of public hearing prior to promulgation of such regulations. In proposing and promulgating regulations under this section, the Administrator shall consult with the Secretary and the National Drinking Water Advisory Council.

(e) The Administrator shall request comments from the Science Advisory Board (established under the Environmental Research, Development, and Demonstration Act of 1978) prior to proposal of a maximum contaminant level goal and national primary drinking water regulation. The Board shall respond, as it deems appropriate, within the time period applicable for promulgation of the national primary drinking water standard concerned. This subsection shall under no circumstances, be used to delay final promulgation of any national primary drinking water standard.

(4) In making the study under this subsection, the National Academy of Sciences (or other organization) shall collect and correlate (A) morbidity and mortality data and (B) monitored data on the quality of drinking water. Any conclusions based on such correlation shall be included in the report of the study.

(5) Neither the report under the study of this subsection nor any draft of such report shall be submitted to the Office of Management and Budget or to any other Federal agency (other than the Environmental Protection Agency) prior to its submission to Congress.

(6) Of the funds authorized to be appropriated to the Administrator by this title, such amounts as may be required shall be available to carry out the study and to make the report directed by paragraph (2) of this subsection.

STATE PRIMARY ENFORCEMENT RESPONSIBILITY

Section 1413. (a) For purposes of this title, a State has primary enforcement responsibility for public water systems during any period for which the Administrator determines (pursuant to regulations prescribed under subsection (b)) that such State—

(1) has adopted drinking water regulations which **are no less stringent than the national primary drinking water regulations in effect under sections 1412(a) and 1412(b);**

(2) has adopted and is implementing adequate procedures for the enforcement of such State regulations, including conducting such monitoring and making such inspections as the Administrator may require by regulation;

(3) will keep such records and make such reports with respect to its activities under paragraphs (1) and (2) as the Administrator may require by regulation;

(4) if it permits variances or exemptions, or both, from the requirements of its drinking water regulations which meet the requirements of paragraph (1), permits such variances and exemptions under conditions and in a matter which is not

less stringent than the conditions under, and the manner in, which variances and exemptions may be granted under sections 1415 and 1416; and

(5) has adopted and can implement an adequate plan for the provision of safe drinking water under emergency circumstances.

(b)(1) The Administrator shall, by regulation (proposed within 180 days of the date of the enactment of this title), prescribe the manner in which a State may apply to the Administrator for a determination that the requirements of paragraphs (1), (2), (3), and (4) of subsection (a) are satisfied with respect to the State, the manner in which the determination is made, the period for which the determination will be effective, and the manner in which the Administrator may determine that such requirements are no longer met. Such regulations shall require that before a determination of the Administrator that such requirements are met or are no longer met with respect to a State may become effective, the Administrator shall notify such State of the determination and the reasons thereof and shall provide an opportunity for public hearing on the determination. Such regulations shall be promulgated (with such modifications as the Administrator deems appropriate) within 90 days of the publication of the proposed regulations in the Federal Register. The Administrator shall promptly notify in writing the chief executive officer of each State of the promulgation of regulations under this paragraph. Such notice shall contain a copy of the regulations and shall specify a State's authority under this title when it is determined to have primary enforcement responsibility for public water systems.

(2) When an application is submitted in accordance with the Administrator's regulations under paragraph (1), the Administrator shall within 90 days of the date on which such application is submitted (A) make the determination applied for, or (B) deny the application and notify the applicant in writing of the reasons for his denial.

ENFORCEMENT OF DRINKING WATER REGULATIONS

Section 1414. (a)(1)(A) Whenever the Administrator finds during a period during which a State has primary enforcement responsibility for public water systems (within the meaning of section 1413 (a)) that any public water system—

(i) for which a variance under section 1415 or an exemption under section 1416 is not in effect, does not comply with any national primary drinking water regulation in effect under section 1412, or

(ii) for which a variance under section 1415 or an exemption under section 1416 is in effect, does not comply with any schedule or other requirement imposed pursuant thereto, he shall so notify the State **and such public water system** and provide such advice and technical assistance to such State and public water system as may be appropriate to bring the system into compliance with such regulation or requirement by the earliest feasible time.

(B) If, beyond the thirtieth day after the Administrator's notification under subparagraph (A), the State has not commenced appropriate enforcement action, the Administrator shall issue an order under subsection (g) requiring

the public water system to comply with such regulation or requirement or the Administrator shall commence a civil action under subsection (b).

(2) Whenever, on the basis of information available to him, the Administrator finds during a period which a State does not have primary enforcement responsibility for public water systems that a public water system in such State—

(A) for which a variance under section 1415(a)(2) or an exemption under section 1416(f) is not in effect, does not comply with any national primary drinking water regulation in effect under section 1412, or

(B) for which a variance under section 1415(a)(2) or an exemption under section 1416(f) is in effect, does not comply with any schedule or other requirement imposed pursuant thereto, **the Administrator shall issue an order under subsection (g) requiring the public water system to comply with such regulation or requirement or the Administrator shall commence a civil action under subsection (b).**

(b) The Administrator may bring a civil action in the appropriate United States district court to require compliance with a national primary drinking water regulation, **with an order issued under subsection (g),** or with any schedule or other requirement imposed pursuant to a variance or exemption granted under section 1415 or 1416 if—

(1) authorized under paragraph (1) or (2) of subsection (a), or

(2) if requested by (A) the chief executive officer of the State in which is located the public water system which is not in compliance with such regulation or requirement, or (B) the agency of such State which has jurisdiction over compliance by public water systems in the State with national primary drinking water regulations or State drinking water regulations.

The court may enter, in an action brought under this subsection, such judgment as protection of public health may require, taking into consideration the time necessary to comply and the availability of alternative water supplies; and, if, the court determines that there has been a violation of the regulation or schedule or other requirement with respect to which the action was brought, the court may, taking into account the seriousness of the violation, the population at risk, and other appropriate factors, impose on the violator a civil penalty of not to exceed **$25,000** for each day in which such violation occurs.

(c) Each owner or operator or a public water system shall give notice to the persons served by it—

(1) of any failure on the part of the public water system to—

(A) comply with an applicable maximum contaminant level or treatment technique requirement of, or a testing procedure prescribed by, a national primary drinking water regulation, or

(B) perform monitoring required by section 1445(a), and

(2) if the public water system is subject to a variance granted under section 1415(a)(1)(A) or 1415(a)(2) for an inability to meet a maximum contaminant level requirement or is subject to an exemption granted under section 1416, of—

(A) the existence of such variance or exemption, and

(B) any failure to comply with the requirements of any schedule prescribed pursuant to the variance or exemption.

The Administrator shall by regulation prescribe the form, manner, and frequency for giving notice under this subsection. **Within 15 months after the enactment of the Safe Drinking Water Act Amendments of 1986, the Administrator shall amend such regulations to provide for different types and frequencies of notice based on the differences between violations which are intermittent or infrequent and violations which are continuous or frequent. Such regulations shall also take into account the seriousness of any potential adverse health effects which may be involved. Notice of any violation of a maximum contaminant level or any other violation designated by the Administrator as posing a serious potential adverse health effect shall be given as soon as possible, but in no case later than 14 days after the violation. Notice of a continuous violation of a regulation other than a maximum contaminant level shall be given no less frequently than every 3 months. Notice of violations judged to be less serious shall be given no less frequently than annually. The Administrator shall specify the types of notice to be used to provide information as promptly and effectively as possible taking into account both the seriousness of any potential adverse health effects and the likelihood of reaching all affected persons. Notification of violations shall include notice by general circulation newspaper serving the area and, whenever appropriate, shall also include a press release to electronic media and individual mailings. Notice under this subsection shall provide a clear and readily understandable explanation of the violation, any potential adverse health effects, the steps that the system is taking to correct such violation, and the necessity for seeking alternative water supplies, if any, until the violation is corrected. Until such amended regulations are promulgated, the regulations in effect on the date of the enactment of the Safe Drinking Water Act Amendments of 1986 shall remain in effect. The Administrator may also require the owner or operator of a public water system to give notice to the persons served by it of contaminant levels of any unregulated contaminant required to be monitored under section 1445(a). Any person who violates this subsection or regulations issued under this subsection shall be subject to a civil penalty of not to exceed $25,000.**

(d) Whenever, on the basis of information available to him, the Administrator finds that within a reasonable time after national secondary drinking water regulations have been promulgated, one or more public water systems in a State do not comply with such secondary regulations, and that such noncompliance appears to result from a failure of such State to take reasonable action to assure that public water systems throughout such State meet such secondary regulations, he shall so notify the State.

(e) Nothing in this title shall diminish any authority of a State or political subdivision to adopt or enforce any law or regulation respecting drinking water regulations or public water systems, but no such law or regulation shall relieve any person of any requirement otherwise applicable under this title.

(f) if the Administrator makes a finding of noncompliance (described in subparagraph (A) or (B) of subsection (a)(1)) with respect to a public water system in a State which has primary enforcement responsibility, the Administrator may, for the purpose of assisting that State in carrying out such responsibility and upon the petition of such State or public water system or persons served by such system, hold, after appropriate notice, public hearings for the purpose of gathering information from technical or other experts, Federal, State, or other public officials, representatives of such public water system, persons served by such system, and other interested persons on—

(1) the ways in which such system can within the earliest feasible time be brought into compliance with the regulation or requirement with respect to which such finding was made, and

(2) the means for the maximum feasible protection of the public health during any period in which such system is not in compliance with a national primary drinking water regulation or requirement applicable to a variance or exemption.

On the basis of such hearings the Administrator shall issue recommendations which shall be sent to such State and public water system and shall be made available to the public and communications media.

(g)(1) In any case in which the Administrator is authorized to bring a civil action under this section or under section 1445 with respect to any regulation, schedule, or other requirement, the Administrator also may issue an order to require compliance with such regulation, schedule, or other requirement.

(2) An order issued under this subsection shall not take effect until after notice and opportunity for public hearing and, in the case of a State having primary enforcement responsibility for public water systems in that State, until after the Administrator has provided the State with an opportunity to confer with the Administrator regarding the proposed order. A copy of any order proposed to be issued under this subsection shall be sent to the appropriate State agency of the State involved if the State has primary enforcement responsibility for public water systems in that State. Any order issued under this subsection shall state with reasonable specificity the nature of the violation. In any case in which an order under this subsection is issued to a corporation, a copy of such order shall be issued to appropriate corporate officers.

(3)(A) Any person who violates, or fails or refuses to comply with, an order under this subsection shall be liable to the United States for a civil penalty of not more than $25,000 per day of violation.

(B) Whenever any civil penalty sought by the Administrator under this paragraph does not exceed a total of $5,000, the penalty shall be assessed by the Administrator after notice and opportunity for a hearing on the record in accordance with section 554 of title 5 of the United States Code.

(C) Whenever any civil penalty sought by the Administrator under this paragraph exceeds $5,000, the penalty shall be assessed by a civil action brought by the Administrator in the appropriate United States district court (as determined under the provisions of title 28 of the United States Code).

(D) If any person fails to pay an assessment of a civil penalty after it has become a final and unappealable order, or after the appropriate court of appeals has entered final judgment in favor of the Administrator, the Attorney General shall recover the amount for which such person is liable in any appropriate district court of the United States. In any such action, the validity and appropriateness of the final order imposing the civil penalty shall not be subject to review.

VARIANCES

Section 1415. (a) Notwithstanding any other provision of this part, variances from national primary drinking water regulations may be granted as follows:

(1)(A) A State which has primary enforcement responsibility for public water systems may grant one or more variances from an applicable national primary drinking water regulation to one or more public water systems within its jurisdiction, which, because of characteristics of the raw water sources which are reasonably available to the systems, cannot meet the requirements respecting the maximum contaminant levels of such drinking water regulation. **A variance may only be issued to a system after the system's** application of the best technology, treatment techniques, or other means, which the Administrator finds are available (taking costs into consideration). **The Administrator shall propose and promulgate his finding of the best available technology, treatment techniques or other means available for each contaminant for purposes of this subsection at the time he proposes and promulgates a maximum contaminant level for each such contaminant. The Administrator's finding of best available technology, treatment techniques or other means for purposes of this subsection may vary depending on the number of persons served by the system or for other physical conditions related to engineering feasibility and costs of compliance with maximum contaminant levels as considered appropriate by the Administrator.** Before a State may grant a variance under this subparagraph, the State must find that the variance will not result in an unreasonable risk to health. If a State grants a public water system a variance under this subparagraph the State shall prescribe **at the time** the variance is granted, a schedule for—

(i) compliance (including increments of progress) by the public water system with each contaminant level requirement with respect to which the variance was granted, and

(ii) implementation by the public water system of such **additional** control measures as the State may require for each contaminant, subject to such contaminant level requirement, during the period ending on the date compliance with such requirement is required.

Before a schedule prescribed by a State pursuant to this subparagraph may take effect, the State shall provide notice and opportunity for a public hearing on the schedule. A notice given pursuant to the preceding sentence may cover the prescribing of more than one such schedule and hearing held pursuant to such notice shall include each of the schedules covered by the notice. A schedule prescribed pursuant to this subparagraph for a public water system granted a variance shall require compliance by the system with each contaminant level requirement with respect to which the variance was granted as expeditiously as practicable (as the State may reasonably determine).

(B) A State which has primary enforcement responsibility for public water systems may grant to one or more public water systems within its jurisdiction one or more variances from any provisions of a national primary drinking water regulation which requires the use of a specified treatment technique with respect to a contaminant if the public water system applying for the variance demonstrates to the satisfaction of the State that such treatment technique is not necessary to protect the health of persons because of the nature of the raw water source of such system. A variance granted under this subparagraph shall be conditioned on such monitoring and other requirements as the Administrator may prescribe.

(C) Before a variance proposed to be granted by a State under subparagraph (A) or (B) may take effect, such State shall provide notice and opportunity for public hearing on the proposed variance. A notice given pursuant to the preceding sentence may cover the granting of more than one variance and a hearing held pursuant to such notice shall include each of the variances covered by the notice. The State shall promptly notify the Administrator of all variances granted by it. Such notification shall contain the reason for the variance (and in the case of a variance under subparagraph (A), the basis for the finding required by that subparagraph before the granting of the variance) and documentation of the need for the variance.

(D) Each public water system's variance granted by a State under subparagraph (A) shall be conditioned by the State upon compliance by the public water system with the schedule prescribed by the State pursuant to that subparagraph. The requirements of each schedule prescribed by a State pursuant to that subparagraph shall be enforceable by the State under its laws. Any requirements of a schedule on which a variance granted under that subparagraph is conditioned may be enforced under section 1414 as if such requirement was part of a national primary drinking water regulation.

(E) Each schedule prescribed by a State pursuant to subparagraph (A) shall be deemed approved by the Administrator unless the variance for which it was prescribed is revoked by the Administrator under such subparagraph.

(F) Not later than 18 months after the effective date of the interim national primary drinking water regulations the Administrator shall complete a comprehensive review of the variances granted under subparagraph (A) (and schedules prescribed pursuant thereto) and under subparagraph (B) by the States during the one-year period beginning on such effective date. The Administrator shall conduct such subsequent review of variances and schedules as he deems necessary to carry out the purposes of this title, but each subsequent review shall be completed within each 3-year period following the completion of the first review under this subparagraph. Before conducting any review under this subparagraph, the Administrator shall publish notice of the proposed review in the Federal Register. Such notice shall (i) provide information respecting the location of data and other information respecting the variances to be reviewed (including data and other information concerning new scientific matters bearing on such variances), and (ii) advise of the opportunity to submit comments on the variances reviewed and on the need for continuing them. Upon completion of any such review, the Administrator shall publish in the Federal Register the results of his review together with findings responsive to comments submitted in connection with such review.

(G)(i) If the Administrator finds that a State has, in a substantial number of instances, abused its discretion in granting variances under subparagraph (A) or (B) or that in a substantial number of cases the State has failed to prescribe schedules in accordance with subparagraph (A), the Administrator shall notify the State of his findings. In determining if a State has abused its discretion in granting variances in a substantial number of instances, the Administrator shall consider the number of persons who are affected by the variances and if the requirements applicable to the granting of the variances were complied with. A notice under this clause shall—

(I) identify each public water system with respect to which the finding was made,

(II) specify the reasons for the finding, and

(III) as appropriate, propose revocations of specific variances or propose revised schedules or other requirements for specific public water systems granted variances, or both.

(ii) The Administrator shall provide reasonable notice and public hearing on the provisions of each notice given pursuant to clause (i) of this subparagraph. After a hearing on a notice pursuant to such clause, the Administrator shall (1) rescind the finding for which the notice was given and promptly notify the State of such rescission, or (II) promulgate (with such modifications as he deems appropriate such variance revocations and revised schedules or other requirements proposed in such notice as he deems appropriate. Not later than 180 days after the date a notice is given pursuant to clause (i) of this subparagraph, the Administrator shall complete the hearing on the notice and take the action required by the preceding sentence.

(iii) If a State is notified under clause (i) of this subparagraph of a finding of the Administrator made with respect to a variance granted a public water system within that State or to a schedule or other requirement for a variance and if, before a revocation of such variance or a revision of such schedule or other requirement promulgated by the Administrator take effect, the State takes corrective action with respect to such variance or schedule or other requirement which the Administrator determines makes his finding inapplicable to such variance or schedule or other requirement, the Administrator shall rescind the application of his finding to that variance or schedule or other requirement. No variance revocation or revised schedule or other requirement may take effect before the expiration of 90 days following the date of the notice in which the revocation or revised schedule or other requirement was proposed.

(2) If a State does not have primary enforcement responsibility for public water systems, the Administrator shall have the same authority to grant variances in such State as the State would have under paragraph (1) if it had primary enforcement responsibility.

(3) The Administrator may grant a variance from any treatment technique requirement of a national primary drinking water regulation upon a showing by any person that an alternative treatment technique not included in such requirement is at least as efficient in lowering the level of the contaminant with respect to which such requirement was prescribed. A variance under this paragraph shall

be conditioned on the use of the alternative treatment technique which is the basis of the variance.

(b) Any schedule or other requirement on which a variance granted under paragraph (1)(B) or (2) of subsection (a) is conditioned may be enforced under section 1414 as if such schedule or other requirement was part of a national primary drinking water regulation.

(c) If an application for variance under subsection (a) is made, the State receiving the application or the Administrator, as the case may be, shall act upon such application within a reasonable period (as determined under regulations prescribed by the Administrator) after the date of its submission.

(d) For purposes of this section, the term "treatment technique requirement" means a requirement in a national primary drinking water regulation which specifies for a contaminant (in accordance with section 1401(1)(C)(ii) each treatment technique known to the Administrator which leads to a reduction in the level of such contaminant sufficient to satisfy the requirements of section 1412(b)(3).

EXEMPTIONS

Section 1416. (a) A State which has primary enforcement responsibility may exempt any public water system within the State's jurisdiction from any requirement respecting a maximum contaminant level or any treatment technique requirement, or from both, of an applicable national primary drinking water regulation upon a finding that—

(1) due to compelling factors (which may include economic factors), the public water system is unable to comply with such contaminant level or treatment technique requirement,

(2) the public water system was in operation on the effective date of such contaminant level or treatment technique requirement or for a system that was not in operation by that date, only if no reasonable alternative source of drinking water is available to such new system, and

(3) the granting of the exemption will not result in an unreasonable risk to health.

(b)(1) If a State grants a public water system an exemption under subsection (a), the State shall prescribe, **at the time** the exemption is granted, a schedule for—

(A) compliance (including increments of progress) by the public water system with each contaminant level requirement and treatment technique requirement with respect to which the exemption was granted, and

(B) implementation by the public water system of such control measures as the State may require for each contaminant, subject to such contaminant level requirement or treatment technique requirement, during the period ending on the date compliance with such requirement is required.

Before a schedule prescribed by a State pursuant to this subsection may take effect, the State shall provide notice and opportunity for a public hearing on the schedule. A notice given pursuant to the preceding sentence may cover the prescribing of more than one such schedule and a hearing held pursuant to such notice shall include each of the schedules covered by the notice.

(2)(A) A schedule prescribed pursuant to this subsection for a public water system granted an exemption under subsection (a) shall require compliance by the system with each contaminant level and treatment technique requirement with respect to which the exemption was granted as expeditiously as practicable (as the State may reasonably determine) but (except as provided in subparagraph (B))—

(i) in the case of an exemption granted with respect to a contaminant level or treatment technique requirement prescribed by the national primary drinking water regulations promulgated under section 1412(a), not later than **12 months after enactment of the SDWA Amendments of 1986; and**

(ii) in the case of an exemption granted with respect to a contaminant level or treatment technique requirement prescribed by national primary drinking water regulations, **other than a regulation referred to in section 1412(a), 12 months after the date of the issuance of the exemption.**

(B) The final date for compliance provided in any schedule in the case of any exemption may be extended by the State (in the case of a State which has primary enforcement responsibility) or by the Administrator (in any other case) for a period not to exceed 3 years after the date of the issuance of the exemption if the public water system establishes that—

(i) the system cannot meet the standard without capital improvements which cannot be completed within the period of such exemption;

(ii) in the case of a system which needs financial assistance for the necessary improvements, the system has entered into an agreement to obtain such financial assistance; or

(iii) the system has entered into an enforceable agreement to become a part of a regional public water system; and the system is taking all practicable steps to meet the standard.

(C) In the case of a system which does not serve more than 500 service connections and which needs financial assistance for the necessary improvements, an exemption granted under clause (i) or (ii) of subparagraph (B) may be renewed for one or more additional 2-year periods if the system establishes that it is taking all practicable steps to meet the requirements of subparagraph (B).

(3) Each public water system's exemption granted by a State under subsection (a) shall be conditioned by the State upon compliance by the public water system with the schedule prescribed by the State pursuant to this subsection. The requirements of each schedule prescribed by a State pursuant to this subsection shall be enforceable by the State under its laws. Any requirement of a schedule on which an exemption granted under this section is conditioned may be enforced under section 1414 as if such requirement was part of a national primary drinking water regulation.

(4) Each schedule prescribed by a State pursuant to this subsection shall be deemed approved by the Administrator unless the exemption for which it was

prescribed is revoked by the Administrator under subsection (d)(2) or the schedule is revised by the Administrator under such subsection.

(c) Each State which grants an exemption under subsection (a) shall promptly notify the Administrator of the granting of such exemption. Such notification shall contain the reasons for the exemption (including the basis for the finding required by subsection (a)(3) before the exemption may be granted) and document the need for the exemption.

(d)(1) Not later than 18 months after the effective date of the interim national primary drinking water regulations the Administrator shall complete a comprehensive review of the exemptions granted (and schedules prescribed pursuant thereto) by the States during the one-year period beginning on such effective date. The Administrator shall conduct such subsequent reviews of exemptions and schedules as he deems necessary to carry out the purposes of this title, but each subsequent review shall be completed within each 3-year period following the completion of the first review under this subparagraph. Before conducting any review under this subparagraph, the Administrator shall publish notice of the proposed review in the Federal Register. Such notice shall (A) provide information respecting the location of data and other information respecting the exemptions to be reviewed (including data and other information concerning new scientific matters bearing on such exemptions), and (B) advise of the opportunity to submit comments on the exemptions reviewed and on the need for continuing them. Upon completion of any such review, the Administrator shall publish in the Federal Register the results of his review together with findings responsive to comments submitted in connection with such review.

(2)(A) If the Administrator finds that a State has, in a substantial number of instances, abused its discretion in granting exemptions under subsection (a) or failed to prescribe schedules in accordance with subsection (b), the Administrator shall notify the State of his finding. In determining if a State has abused its discretion in granting exemptions in a substantial number of instances, the Administrator shall consider the number of persons who are affected by the exemptions and if the requirements applicable to the granting of the exemptions were complied with. A notice under this paragraph shall—

(i) identify each exempt public water system with respect to which the finding was made,

(ii) specify the reasons for the finding, and

(iii) as appropriate, propose revocations of specific exemptions or propose revised schedules for specific exempt public water systems, or both.

(B) The Administrator shall provide reasonable notice and public hearing on the provisions of each notice given pursuant to subparagraph (A). After a hearing on a notice pursuant to subparagraph (A), the Administrator shall (i) rescind the finding for which the notice was given and promptly notify the State of such rescission, or (ii) promulgate (with such modifications as he deems appropriate) such exemption revocations and revised schedules proposed in such notice as he deems appropriate. Not later than 180 days after the date a notice is given pursuant to subparagraph (A), the Administrator shall complete the hearing on the notice and take the action required by the preceding sentence.

(C) If a State is notified under subparagraph (A) of a finding of the Administrator made with respect to an exemption granted a public water system within that State or to a schedule prescribed pursuant to such an exemption and if before a revocation of such exemption or a revision of such schedule promulgated by the Administrator takes effect the State takes corrective action with respect to such exemption or schedule which the Administrator determines makes his finding inapplicable to such exemption or schedule, the Administrator shall rescind the application of his finding to that exemption or schedule. No exemption revocation or revised schedule may take effect before the expiration of 90 days following the date of the notice in which the revocation or revised schedule was proposed.

(e) For purposes of this section, the term "treatment technique requirement" means a requirement in a national primary drinking water regulation which specifies for a contaminant (in accordance with section 1401 (1)(C)(ii) each treatment technique known to the Administrator which leads to a reduction in the level of such contaminant sufficient to satisfy the requirements of section 1412(b).

(f) If a State does not have primary enforcement responsibility for public water systems, the Administrator shall have the same authority to exempt public water systems in such State from maximum contaminant level requirements and treatment technique requirements under the same conditions and in the same manner as the State would be authorized to grant exemptions under this section if it had primary enforcement responsibility.

(g) If an application for an exemption under this section is made, the State receiving the application or the Administrator, as the case may be, shall act upon such application within a reasonable period (as determined under regulations prescribed by the Administrator) after the date of its submission.

Section 1417. PROHIBITION ON USE OF LEAD PIPES, SOLDER, AND FLUX

(a) IN GENERAL—

(1) PROHIBITION—Any pipe, solder, or flux, which is used after the enactment of the Safe Drinking Water Act Amendments of 1986, in the installation or repair of—

(A) any public water system, or

(B) any plumbing in a residential or nonresidential facility providing water for human consumption which is connected to a public water system, shall be lead free (within the meaning of subsection (d)). This paragraph shall not apply to leaded joints necessary for the repair of cast iron pipes.

(2) PUBLIC NOTICE REQUIREMENTS

(A) IN GENERAL—Each public water system shall identify and provide notice to persons that may be affected by lead contamination of their drinking water where such contamination results from either or both of the following:

(i) The lead content in the construction materials of the public water distribution system.

(ii) Corrosivity of the water supply sufficient to cause leaching of lead. The notice shall be provided in such manner and form as may be reasonably required by the Administrator. Notice under this paragraph shall be provided notwithstanding the absence of a violation of any national drinking water standard.

(B) CONTENTS OF NOTICE—Notice under this paragraph shall provide a clear and readily understandable explanation of—

(i) the potential sources of lead in the drinking water,

(ii) potential adverse health effects,

(iii) reasonably available methods of mitigating known or potential lead content in drinking water,

(iv) any steps the system is taking to mitigate lead content in drinking water, and

(v) the necessity for seeking alternative water supplies, if any.

(b) STATE ENFORCEMENT—

(1) ENFORCEMENT OF PROHIBITION—The requirements of subsection (a)(1) shall be enforced in all States effective 24 months after the enactment of this section. States shall enforce such requirements through State of local plumbing codes, or such other means of enforcement as the State may determine to be appropriate.

(2) ENFORCEMENT OF PUBLIC NOTICE REQUIREMENTS—The requirements of subsection (a)(2) shall apply to all States effective 24 months after the enactment of this section.

(c) PENALTIES—If the Administrator determines that a State is not enforcing the requirements of subsection (a) as required pursuant to subsection (b), the Administrator may withhold up to 5 percent of Federal funds available to that State for State program grants under section 1443(a).

(d) DEFINITION OF LEAD FREE—For purposes of this section, the term "lead free"—

(1) when used with respect to solders and flux refers to solders and flux containing not more than 0.2 percent lead, and

(2) when used with respect to pipes and pipe fittings refers to pipes and pipe fittings containing not more than 8.0 percent lead.

Part C—Protection of Underground Sources of Drinking Water

REGULATIONS FOR STATE PROGRAMS

Section 1421. (a)(1) The Administrator shall publish proposed regulations for State underground injection control programs within 180 days after the date of enactment of this title. Within 180 days after publication of such proposed regulations, he shall promulgate such regulations with such modifications as he deems appropriate. Any regulation under this subsection may be amended from time to time.

(2) Any regulation under this section shall be proposed and promulgated in accordance with section 553 of title 5, United States Code (relating to rulemaking), except that the Administrator shall provide opportunity for public hearing

prior to promulgation of such regulations. In proposing and promulgating regulations under this section, the Administrator shall consult with the Secretary, the National Drinking Water Advisory Council, and other appropriate Federal entities and with interested State entities.

(b)(1) Regulations under subsection (a) for State underground injection programs shall contain minimum requirements for effective programs to prevent underground injection which endangers drinking water sources within the meaning of subsection (d)(2). Such regulations shall require that a State program, in order to be approved under section 1422—

(A) shall prohibit, effective on the date on which the applicable underground injection control program takes effect, any underground injection in such State which is not authorized by a permit issued by the State (except that the regulations may permit a State to authorize underground injection by rule);

(B) shall require (i) in the case of a program which provides for authorization of underground injection by permit, that the applicant for the permit to inject must satisfy the State that the underground injection will not endanger drinking water sources, and (ii) in the case of a program which provides for such an authorization by rule, that no rule may be promulgated which authorizes any underground injection which endangers drinking water sources;

(C) shall include inspection, monitoring, recordkeeping, and reporting requirements; and

(D) shall apply (i) as prescribed by section 1447(b), to underground injections by Federal agencies, and (ii) to underground injections by any other person whether or not occurring on property owned or leased by the United States.

(2) Regulations of the Administrator under this section for State underground injection control programs may not prescribe requirements which interfere with or impede—

(A) the underground injection of brine or other fluids which are brought to the surface in connection with oil or natural gas production **or natural gas storage requirements**, or

(B) any underground injection for the secondary or tertiary recovery of oil or natural gas,

unless such requirements are essential to assure that underground sources of drinking water will not be endangered by such injection.

(3)(A) The regulations of the Administrator under this section shall permit or provide for consideration of varying geologic, hydrological, or historical conditions in different States and in different areas within a State.

(B)(i) In prescribing regulations under this section the Administrator shall, to the extent feasible, avoid promulgation of requirements which would unnecessarily disrupt State underground injection control programs which are in effect and being enforced in a substantial number of States.

(ii) For the purpose of this subparagraph, a regulation prescribed by the Administrator under this section shall be deemed to disrupt a State underground injection control program only if it would be infeasible to comply with both such regulation and the State underground injection control program.

(iii) For the purpose of this subparagraph, a regulation prescribed by the Administrator under this section shall be deemed unnecessary only if, without such regulation, underground sources of drinking water will not be endangered by any underground injection.

(C) Nothing in this section shall be construed to alter or affect the duty to assure that underground sources of drinking water will not be endangered by any underground injection.

(c)(1) The Administrator may, upon application of the Governor of a State which authorizes underground injection by means of permits, authorize such State to issue (without regard to subsection (b)(1)(B)(i)) temporary permits for underground injection which may be effective until the expiration of four years after the date of enactment of this title, if—

(A) the Administrator finds that the State has demonstrated that it is unable and could not reasonably have been able to process all permit applications within the time available;

(B) the Administrator determines the adverse effect on the environment of such temporary permits is not unwarranted;

(C) such temporary permits will be issued only with respect to injection wells in operation on the date on which such State's permit program approved under this part first takes effect and for which there was inadequate time to process its permit application; and

(D) the Administrator determines the temporary permits require the use of adequate safeguards established by rules adopted by him.

(2) The Administrator may, upon application of the Governor of a State which authorizes underground injection by means of permits, authorize such State to issue (without regard to subsection (b)(1)(B)(i)), but after reasonable notice and hearing, one or more temporary permits each of which is applicable to a particular injection well and to the underground injection of a particular fluid and which may be effective until the expiration of four years after the date of enactment of this title, if the State finds, on the record of such hearing—

(A) that technology (or other means) to permit safe injection of the fluid in accordance with the applicable underground injection control program is not generally available (taking costs into consideration);

(B) that injection of the fluid would be less harmful to health than the use of other available means of disposing of waste or producing the desired product; and

(C) that available technology or other means have been employed (and will be employed) to reduce the volume and toxicity of the fluid and to minimize the potentially adverse effect of the injection on the public health.

(d) For purposes of this part:

(1) The term "underground injection" means the subsurface emplacement of fluids by well injection. Such term does not include the underground injection of natural gas for purposes of storage.

(2) Underground injection endangers drinking water sources if such injection

may result in the presence in underground water which supplies or can reasonably be expected to supply any public water system of any contaminant, and if the presence of such contaminant may result in such system's not complying with any national primary drinking water regulation or may otherwise adversely affect the health of persons.

STATE PRIMARY ENFORCEMENT RESPONSIBILITY

Section 1422. (a) Within 180 days after the date of enactment of this title, the Administrator shall list in the Federal Register each State for which in his judgment a State underground injection control program may be necessary to assure that underground injection will not endanger drinking water sources. Such list may be amended from time to time.

(b)(1)(A) Each State listed under subsection (a) shall within 270 days after the date of promulgation of any regulation under section 1421 (or, if later, within 270 days after such State is first listed under subsection (a)) submit to the Administrator an application which contains a showing satisfactory to the Administrator that the State—

(i) has adopted after reasonable notice and public hearings, and will implement, an underground injection control program which meets the requirements of regulations in effect under section 1421; and

(ii) will keep such records and make such reports with respect to its activities under its underground injection control program as the Administrator may require by regulation.

The Administrator may, for good cause, extend the date for submission of an application by any State under this subparagraph for a period not to exceed an additional 270 days.

(B) Within 270 days of any amendment of a regulation under section 1421 revising or adding any requirement respecting State underground injection control programs, each State listed under subsection (a) shall submit (in such form and manner as the Administrator may require) a notice to the Administrator containing a showing satisfactory to him that the State underground injection control program meets the revised or added requirement.

(2) Within ninety days after the State's application under paragraph (1)(A) or notice under paragraph (1)(B) and after reasonable opportunity for presentation of views, the Administrator shall by rule either approve, disapprove, or approve in part and disapprove in part, the State's underground injection control program.

(3) If the Administrator approves the State's program under paragraph (2), the State shall have primary enforcement responsibility for underground water sources until such time as the Administrator determines, by rule, that such State no longer meets the requirements of clause (i) or (ii) of paragraph (1)(A) of this subsection.

(4) Before promulgating any rule under paragraph (2) or (3) of this subsection, the Administrator shall provide opportunity for public hearing respecting such rule.

(c) If the Administrator disapproves a State's program (or part thereof) under subsection (b)(2), if the Administrator determines under subsection (b)(3) that a State no longer meets the requirements of clause (i) or (ii) of subsection (b)(1)(A), or if a State fails to submit an application or notice before the date of expiration of the period specified in subsection (b)(1), the Administrator shall by regulation within 90 days after the date of such disapproval, determination, or expiration (as the case may be) prescribe (and may from time to time by regulation revise) a program applicable to such State meeting the requirements of section 1421(b). Such program may not include requirements which interfere with or impede—

(1) the underground injection of brine or other fluids which are brought to the surface in connection with oil or natural gas production **or natural gas storage operations,** or

(2) any underground injection for the secondary or tertiary recovery of oil or natural gas,

unless such requirements are essential to assure that underground sources of drinking water will not be endangered by such injection. Such program shall apply in such State to the extent that a program adopted by such State which the Administrator determines meets such requirements is not in effect. Before promulgating any regulation under this section, the Administrator shall provide opportunity for public hearing respecting such regulation.

(d) For purposes of this title, the term "applicable underground injection control program" with respect to a State means the program (or most recent amendment thereof) (1) which has been adopted by the State and which has been approved under subsection (b), or (2) which has been prescribed by the Administrator under subsection (c).

(e) An Indian Tribe may assume primary enforcement responsibility for underground injection control under this section consistent with such regulations as the Administrator has prescribed pursuant to Part C and section 1451 of this Act. The area over which such Indian Tribe exercises governmental jurisdiction need not have been listed under subsection (a) of this section, and such Tribe need not submit an application to assume primary enforcement responsibility within the 270-day deadline noted in subsection (b)(1)(A) of this section. Until an Indian Tribe assumes primary enforcement responsibility, the currently applicable underground injection control program shall continue to apply. If an applicable underground injection control program does not exist for an Indian Tribe, the Administrator shall prescribe such a program pursuant to subsection (c) of this section, and consistent with section 1412 (b), within 270 days after the enactment of the Safe Drinking Water Act Amendments of 1986, unless an Indian Tribe first obtains approval to assume primary enforcement responsibility for underground injection control.

ENFORCEMENT OF PROGRAM

Section 1423. (a)(1) Whenever the Administrator finds during a period during which a State has primary enforcement responsibility for underground water sources (within the meaning of section 1422(b)(3) or section 1425 (c)) that any person who is subject to a requirement of an applicable underground injection control program in such State is violating such requirement, he shall so notify the State and the person violating such requirement. **If beyond the thirtieth day after the Administrator's notification the State has not commenced appropriate enforcement action, the Administrator shall issue an order under subsection (c) requiring the person to comply with such requirement or the Administrator shall commence a civil action under subsection (b).**

(2) Whenever the Administrator finds during a period during which a State does not have primary enforcement responsibility for underground water sources that any person subject to any requirement of any applicable underground injection control program in such States is violating such requirement, **the Administrator shall issue an order under subsection (c) requiring the person to comply with such requirement or the Administrator shall commence a civil action under subsection (b).**

(b) CIVIL AND CRIMINAL ACTIONS—Civil actions referred to in paragraphs (1) and (2) of subsection (a) shall be brought in the appropriate United States district court. Such court shall have jurisdiction to require compliance with any requirement of an applicable underground injection program or with an order issued under subsection (c). The court may enter such judgment as protection of public health may require. Any person who violates any requirement of an applicable underground injection control program or an order requiring compliance under subsection (c)—

(1) shall be subject to a civil penalty of not more than $25,000 for each day of such violation, and

(2) if such violation is willful, such person may, in addition to or in lieu of the civil penalty authorized by paragraph (1), be imprisoned for not more than 3 years, or fined in accordance with title 18 of the United States Code, or both.

(c) ADMINISTRATIVE ORDERS—(1) In any case in which the Administrator is authorized to bring a civil action under this section with respect to any regulation or other requirement of this part other than those relating to—

(A) the underground injection of brine or other fluids which are brought to the surface in connection with oil or natural gas production, or

(B) any underground injection for the secondary or tertiary recovery of oil or natural gas, the Administrator may also issue an order under this subsection either assessing a civil penalty of not more than $10,000 for each day of violation for any past or current violation, up to a maximum administrative penalty of $125,000, or requiring compliance with such regulation or other requirement, or both.

(2) In any case in which the Administrator is authorized to bring a civil action under this section with respect to any regulation, or other requirement of this part relating to—

(A) the underground injection of brine or other fluids which are brought to the surface in connection with oil or natural gas production, or

(B) any underground injection for the secondary or tertiary recovery of oil or natural gas,

the Administrator may also issue an order under this subsection either assessing a civil penalty of not more than $5,000 for each day of violation for any past or

current violation, up to a maximum administrative penalty of $125,000, or requiring compliance with such regulation or other requirement, or both.

(3)(A) An order under this subsection shall be issued by the Administrator after opportunity (provided in accordance with this subparagraph) for a hearing. Before issuing the order, the Administrator shall give to the person to whom it is directed written notice of the Administrator's proposal to issue such order and the opportunity to request, within 30 days of the date the notice is received by such person, a hearing on the order. Such hearing shall not be subject to section 554 or 556 of title 5, United States Code, but shall provide a reasonable opportunity to be heard and to present evidence.

(B) The Administrator shall provide public notice of, and reasonable opportunity to comment on, any proposed order.

(C) Any citizen who comments on any proposed order under subparagraph (B) shall be given notice of any hearing under this subsection and of any order. In any hearing held under subparagraph (A), such citizen shall have a reasonable opportunity to be heard and to present evidence.

(D) Any order issued under this subsection shall become effective 30 days following its issuance unless an appeal is taken pursuant to paragraph (6).

(4)(A) Any order issued under this subsection shall state with reasonable specificity the nature of the violation and may specify a reasonable time for compliance.

(B) In assessing any civil penalty under this subsection, the Administrator shall take into account appropriate factors, including (i) the seriousness of the violation; (ii) the economic benefit (if any) resulting from the violation; (iii) any history of such violations; (iv) any good-faith efforts to comply with the applicable requirements; (v) the economic impact of the penalty on the violator; and (vi) such other matters as justice may require.

(5) Any violation with respect to which the Administrator has commenced and is diligently prosecuting an action, or has issued an order under this subsection assessing a penalty, shall not be subject to an action under subsection (b) of this section or section 1424(c) or 1449, except that the foregoing limitation on civil actions under section 1449 of this Act shall not apply with respect to any violation for which—

(A) a civil action under section 1449(a)(1) has been filed prior to commencement of an action under this subsection, or

(B) a notice of violation under section 1449(b)(I) has been given before commencement of an action under this subsection and an action under section 1449(a)(1) of this Act is filed before 120 days after such notice is given.

(6) Any person against whom an order is issued or who commented on a proposed order pursuant to paragraph (3) may file an appeal of such order with the United States District Court for the District of Columbia or the district in which the violation is alleged to have occurred. Such an appeal may only be filed within the 30-day period beginning on the date the order is issued. Appellant shall simultaneously send a copy of the appeal by certified mail to the Administrator and to the Attorney General. The Administrator shall promptly file in such court a certified copy of the record on which such order was imposed. The district

court shall not set aside or remand such order unless there is not substantial evidence on the record, taken as a whole, to support the finding of a violation or, unless the Administrator's assessment of penalty or requirement for compliance constitutes an abuse of discretion. The district court shall not impose additional civil penalties for the same violation unless the Administrator's assessment of a penalty constitutes an abuse of discretion. Notwithstanding section 1448(a)(2), any order issued under paragraph (3) shall be subject to judicial review exclusively under this paragraph.

(7) If any person fails to pay an assessment of a civil penalty—

(A) after the order becomes effective under paragraph (3), or

(B) after a court, in an action brought under paragraph (6), has entered a final judgment in favor of the Administrator,

the Administrator may request the Attorney General to bring a civil action in an appropriate district court to recover the amount assessed (plus costs, attorneys' fees, and interest at currently prevailing rates from the date the order is effective or the date of such final judgment, as the case may be). In such an action, the validity, amount, and appropriateness of such penalty shall not be subject to review.

(8) The Administrator may, in connection with administrative proceedings under this subsection, issue subpoenas compelling the attendance and testimony of witnesses and subpoenas duces tecum, and may request the Attorney General to bring an action to enforce any subpoena under this section. The district courts shall have jurisdiction to enforce such subpoenas and impose sanction.

(d) Nothing in this title shall diminish any authority of a State or political subdivision to adopt or enforce any law or regulation respecting underground injection but no such law or regulation shall relieve any person of any requirement otherwise applicable under this title.

INTERIM REGULATION OF UNDERGROUND INJECTIONS

Section 1424. (a)(1) Any person may petition the Administrator to have an area of a State (or States) designated as an area in which no new underground injection well may be operated during the period beginning on the date of the designation and ending on the date on which the applicable underground injection control program covering such area takes effect unless a permit for the operation of such well has been issued by the Administrator under subsection (b). The Administrator may so designate an area within a state if he finds that the area has one aquifer which is the sole or principal drinking water source for the area and which, if contaminated, would create a significant hazard to public health.

(2) Upon receipt of a petition under paragraph (1) of this subsection, the Administrator shall publish it in the Federal Register and shall provide an opportunity to interested persons to submit written data, views, or arguments thereon. Not later than the 30th day following the date of the publication of a petition under this paragraph in the Federal Register, the Administrator shall either make the designation for which the petition is submitted or deny the petition.

(b)(1) During the period beginning on the date an area is designated under subsection (a) and ending on the date the applicable underground injection control program covering such area takes effect, no new underground injection well may be operated in such area unless the Administrator has issued a permit for such operation.

(2) Any person may petition the Administrator for the issuance of a permit for the operation of such a well in such an area. A petition submitted under this paragraph shall be submitted in such manner and contain such information as the Administrator may require by regulation. Upon receipt of such a petition, the Administrator shall publish it in the Federal Register. The Administrator shall give notice of any proceeding on a petition and shall provide opportunity for agency hearing. The Administrator shall act upon such petition on the record of any hearing held pursuant to the preceding sentence respecting such petition. Within 120 days of the publication in the Federal Register of a petition submitted under this paragraph, the Administrator shall either issue the permit for which the petition was submitted or shall deny its issuance.

(3) The Administrator may issue a permit for the operation of a new underground injection well in an area designated under subsection (a) only if he finds that the operation of such well will not cause contamination of the aquifer of such area so as to create a significant hazard to public health. The Administrator may condition the issuance of such a permit upon the use of such control measures in connection with the operation of such well, for which the permit is to be issued, as he deems necessary to assure that the operation of the well will not contaminate the aquifer of the designated area in which the well is located so as to create a significant hazard to public health.

(c) Any person who operates a new underground injection well in violation of subsection (b), (1) shall be subject to a civil penalty of not more than $5,000 for each day in which such violation occurs, or (2) if such violation is willful, such person may, in lieu of the civil penalty authorized by clause (1) be fined not more than $10,000 for each day in which such violation occurs. If the Administrator has reason to believe that any person is violating or will violate subsection (b), he may petition the United States district court to issue a temporary restraining order or injunction (including a mandatory injunction) to enforce such subsection.

(d) For purposes of this section, the term "new underground injection well" means an underground injection well whose operation was not approved by appropriate State and Federal agencies before the date of the enactment of this title.

(e) If the Administrator determines, on his own initiative or upon petition, that an area has an aquifer which is the sole or principal drinking water source for the area and which, if contaminated, would create a significant hazard to public health, he shall publish notice of that determination in the Federal Register. After the publication of any such notice, no commitment for Federal financial assistance (through a grant, contract, loan guarantee, or otherwise) may be entered into for any project which the Administrator determines may contaminate such aquifer through a recharge zone so as to create a significant hazard to public health, but a commitment for Federal financial assistance may, if authorized under another provision of law, be entered into a plan or design the project to assure that it will not so contaminate the aquifer.

OPTIONAL DEMONSTRATION BY STATES RELATING TO OIL OR NATURAL GAS

Section 1425. (a) For purposes of the Administrator's approval or disapproval under section 1422 of that portion of any State underground injection control program which relates to—

(1) the underground injection of brine or other fluids which are brought to the surface in connection with oil or natural gas production **or natural gas storage operation,** or

(2) any underground injection for the secondary or tertiary recovery of oil or natural gas,

in lieu of the showing required under subparagraph (A) of section 1422(b)(1) the State may demonstrate that such portion of the State program meets the requirements of subparagraphs (A) through (D) of section 1421(b)(1) and represents an effective program (including adequate recordkeeping and reporting) to prevent underground injection which endangers drinking water sources.

(b) if the Administrator revises or amends any requirement of a regulation under section 1421 relating to any aspect of the underground injection referred to in subsection (a), in the case of that portion of a State underground injection control program for which the demonstration referred to in subsection (a) has been made, in lieu of the showing required under section 1422(b)(1)(B) the State may demonstrate that, with respect to that aspect of such underground injection, the State program meets the requirements of subparagraphs (A) through (D) of section 1421(b)(1) and represents an effective program (including adequate recordkeeping and reporting) to prevent underground injection which endangers drinking water sources.

(c)(1) Section 1422(b)(3) shall not apply to that portion of any State underground injection control program approved by the Administrator pursuant to a demonstration under subsection (a) of this section (and under subsection (b) of this section where applicable).

(2) If pursuant to such a demonstration, the Administrator approves such portion of the State program, the State shall have primary enforcement responsibility with respect to that portion until such time as the Administrator determines, by rule, that such demonstration is no longer valid. Following such a determination, the Administrator may exercise the authority of subsection (c) of section 1422 in the same manner as provided in such subsection with respect to a determination described in such subsections.

(3) Before promulgating any rule under paragraph (2), the Administrator shall provide opportunity for public hearing respecting such rule.

Section 1426. REGULATION OF STATE PROGRAMS

(a) MONITORING METHODS—Not later than 18 months after enactment of the Safe Drinking Water Act Amendments of 1986, the Administrator shall modify regulations issued under this Act for Class I injection wells to identify

monitoring methods, in addition to those in effect on November 1, 1985, including groundwater monitoring. In accordance with such regulations, the Administrator, or delegated State authority, shall determine the applicability of such monitoring methods, wherever appropriate, at locations and in such a manner as to provide the earliest possible detection of fluid migration into, or in the direction of, underground sources of drinking water from such wells, based on its assessment of the potential for fluid migration from the injection zone that may be harmful to human health or the environment. For purposes of this subsection, a class I injection well is defined in accordance with 40 CFR 146.05 as in effect on November 1, 1985.

(b) REPORT—The Administrator shall submit a report to Congress, no later than September 1987, summarizing the results of State surveys required by the Administrator under this section. The report shall include each of the following items of information:

(1) The numbers and categories of class V wells which discharge nonhazardous waste into or above an underground source of drinking water.

(2) The primary contamination problems associated with different categories of these disposal wells.

(3) Recommendations for minimum design, construction, installation, and siting requirements that should be applied to protect underground sources of drinking water from such contamination wherever necessary.

Section 1427. SOLE SOURCE AQUIFER DEMONSTRATION PROGRAM

(a) PURPOSE—The purpose of this section is to establish procedures for development, implementation, and assessment of demonstration programs designed to protect critical aquifer protection areas located within areas designated as sole or principal source aquifers under section 1424(e) of this Act.

(b) DEFINITION—For purposes of this section, the term "critical aquifer protection area" means either of the following:

(1) All of part of an area located within an area for which an application or designation as a sole or principal source aquifer pursuant to section 1424(e), has been submitted and approved by the Administrator not later than 24 months after the enactment of the Safe Drinking Water Act Amendments of 1986 and which satisfies the criteria established by the Administrator under subsection (d).

(2) All or part of an area which is within an aquifer designated as a sole source aquifer as of the enactment of the Safe Drinking Water Act Amendments of 1986 and for which an areawide ground water quality protection plan has been approved under section 208 of the Clean Water Act prior to such enactment.

(c) APPLICATION—Any State, municipal or local government or political subdivision thereof or any planning entity (including any interstate regional planning entity) that identifies a critical aquifer protection area over which it has authority or jurisdiction may apply to the Administrator for the selection of such area for a demonstration program under this section. Any applicant shall consult with other government or planning entities with authority or jurisdiction in such area prior to application. Applicants, other than the Governor, shall submit the application for a demonstration program jointly with the Governor.

(d) CRITERIA—Not later than 1 year after the enactment of the Safe Drinking Water Act Amendments of 1986, the Administrator shall, by rule, establish criteria for identifying critical aquifer protection areas under this section. In establishing such criteria, the Administrator shall consider each of the following:

(1) The vulnerability of the aquifer to contamination due to hydrogeologic characteristics.

(2) The number of persons or the portion of population using the ground water as a drinking water source.

(3) The economic, social and environmental benefits that would result to the area from maintenance of ground water of high quality.

(4) The economic, social and environmental benefits that would result from degradation of the quality of the ground water.

(e) CONTENTS OF APPLICATION—An application submitted to the Administrator by any applicant for a demonstration program under this section shall meet each of the following requirements:

(1) The application shall propose boundaries for the critical aquifer protection area within its jurisdiction.

(2) The application shall designate or, if necessary, establish a planning entity (which shall be a public agency and which shall include representation of elected local and State governmental officials) to develop a comprehensive management plan (hereinafter in this section referred to as the "plan") for the critical protection area. Where a local government planning agency exists with adequate authority to carry out this section with respect to any proposed critical protection area, such agency shall be designated as the planning entity.

(3) The application shall establish procedures for public participation in the development of the plan, for review, approval, and adoption of the plan, and for assistance to municipalities and other public agencies with authority under State law to implement the plan.

(4) The application shall include a hydrogeologic assessment of surface and ground water resources within the critical protection area.

(5) The application shall include a comprehensive management plan for the proposed protection area.

(6) The application shall include the measures and schedule proposed for implementation of such plan.

(f) COMPREHENSIVE PLAN—

(1) The objective of a comprehensive management plan submitted by an applicant under this section shall be to maintain the quality of the ground water in the critical protection area in a manner reasonably expected to protect human health, the environment and ground water resources. In order to achieve such objective, the plan may be designed to maintain, to the maximum extent possible, the natural vegetative and hydrogeological conditions. Each of the following elements shall be included in such a protection plan:

(A) A map showing the detailed boundary of the critical protection area.

(B) An identification of existing and potential point and nonpoint sources of ground water degradation.

(C) An assessment of the relationship between activities on the land surface and ground water quality.

(D) Specific actions and management practice to be implemented in the critical protection area to prevent adverse impacts on ground water quality.

(E) Identification of authority adequate to implement the plan, estimates of program costs, and sources of State matching funds.

(2) Such plan may also include the following:

(A) A determination of the quality of the existing ground water recharged through the special protection area and the natural recharge capabilities of the special protection area watershed.

(B) Requirements designed to maintain existing underground drinking water quality or improve underground drinking water quality if prevailing conditions fail to meet drinking water standards, pursuant to this Act and State law.

(C) Limits on Federal, State, and local government, financially assisted activities and projects which may contribute to degradation of such ground water or any loss of natural surface and subsurface infiltration of purification capability of the special protection watershed.

(D) A comprehensive statement of land use management including emergency contingency planning as it pertains to the maintenance of the quality of underground sources of drinking water or to the improvement of such sources if necessary to meet drinking water standards pursuant to this Act and State law.

(E) Actions in the special protection area which would avoid adverse impacts on water quality, recharge capabilities, or both.

(F) Consideration of specific techniques, which may include clustering, transfer of development rights, and other innovative measures sufficient to achieve the objectives of this section.

(G) Consideration of the establishment of a State institution to facilitate and assist funding a development transfer credit system.

(H) A program for State and local implementation of the plan described in this subsection in a manner that will insure the continued, uniform, consistent protection of the critical protection area in accord with the purposes of this section.

(I) Pollution abatement measures, if appropriate.

(g) PLANS UNDER SECTION 208 OF THE CLEAN WATER ACT—A plan approved before the enactment of the Safe Drinking Water Act Amendment of 1986 under section 208 of the Clean Water Act to protect a sole source aquifer designated under section 1424(e) of this Act shall be considered a comprehensive management plan for the purposes of this section.

(h) CONSULTATION AND HEARINGS—During the development of a comprehensive management plan under this section, the planning entity shall consult with, and consider the comments of, appropriate officials of any municipality and State or Federal agency which has jurisdiction over lands and waters within the special protection area, other concerned organizations and technical and citizen advisory committees. The planning entity shall conduct public hearings at

places within the special protection area for the purpose of providing the opportunity to comment on any aspect of the plan.

(i) APPROVAL OR DISAPPROVAL—Within 120 days after receipt of an application under this section, the Administrator shall approve or disapprove the application. The approval or disapproval shall be based on a determination that the critical protection area satisfies the criteria established under subsection (d) and that a demonstration program for the area would provide protection for ground water quality consistent with the objective stated in subsection (f) The Administrator shall provide to the Governor a written explanation of the reasons for the disapproval of any such application. Any petitioner may modify and resubmit any application which is not approved. Upon approval of an application, the Administrator may enter into a cooperative agreement with the applicant to establish a demonstration program under this section.

(j) GRANTS AND REIMBURSEMENT—Upon entering a cooperative agreement under subsection (i), the Administrator may provide to the applicant, on a matching basis, a grant of 50 per centum of the costs of implementing the plan established under this section. The Administrator may also reimburse the applicant of an approved plan up to 50 centum of the costs of developing such plan, except for plan approved under section 208 of the Clean Water Act. The total amount of grants under this section for any one aquifer, designated under section 1424(e), shall not exceed $4,000,000 in any one fiscal year.

(k) ACTIVITIES FUNDED UNDER OTHER LAW—No funds authorized under this subsection may be used to fund activities funded under other sections of this Act or the Clean Water Act, the Solid Waste Disposal Act, the Comprehensive Environmental Response, Compensation, and Liability Act of 1980 or other environmental laws.

(l) REPORT—Not later than December 31, 1989, each State shall submit to the Administrator a report assessing the impact of the program on ground water quality and identifying those measures found to be effective in protecting ground water resources. No later than September 30, 1990, the Administrator shall submit to Congress a report summarizing the State reports, and assessing the accomplishments of the sole source aquifer demonstration program including an identification of protection methods found to be most effective and recommendations for their application to protect ground water resources from contamination whenever necessary.

(m) SAVINGS PROVISION—Nothing under this section shall be construed to amend, supersede or abrogate rights to quantities of water which have been established by interstate water compacts, Supreme Court decrees, or State water laws; or any requirement imposed or right provided under any Federal or State environmental or public health statute.

(n) AUTHORIZATION—There are authorized to be appropriated to carry out this section not more than the following amounts:

Fiscal year:	Amount
1987	$10,000,000
1988	15,000,000
1989	17,500,000

1990--17,500,000
1991--17,500,000

Matching grants under this section may also be used to implement or update any water quality management plan for a sole or principal source aquifer approved (before the date of the enactment of this section) by the Administrator under section 208 of the Federal Water Pollution Control Act.

Section 1428. STATE PROGRAMS TO ESTABLISH WELLHEAD PROTECTION AREAS

(a) STATE PROGRAMS—The Governor or Governor's designee of each State shall, within 3 years of the date of enactment of the Safe Drinking Water Act Amendments of 1986, adopt and submit to the Administrator a State program to protect wellhead areas within their jurisdiction from contaminants which may have any adverse affects on the health of persons. Each State program under this section shall, at a minimum—

(1) specify the duties of State agencies, local governmental entities, and public waste supply systems with respect to the development and implementation of programs required by this section;

(2) for each wellhead, determine the wellhead protection area as defined in subsection (e) based on all reasonably available hydrogeologic information on ground water flow, recharge and discharge and other information the State deems necessary to adequately determine the wellhead protection area;

(3) identify within each wellhead protection area all potential anthropogenic sources of contaminants which may have any adverse effect on the health of persons;

(4) describe a program that contains, as appropriate, technical assistance, financial assistance, implementation of control measures, education, training, and demonstration projects to protect the water supply within wellhead protection areas from such contaminants;

(5) include contingency plans for the location and provision of alternate drinking water supplies for each public water system in the event of well or wellfield contamination by such contaminants; and

(6) include a requirement that consideration be given to all potential sources of such contaminants within the expected wellhead area of a new water well which serves a public water supply system.

(b) PUBLIC PARTICIPATION—To the maximum extent possible, each State shall establish procedures, including but not limited to the establishment of technical and citizens' advisory committees, to encourage the public to participate in developing the protection program for wellhead areas. Such procedures shall include notice and opportunity for public hearing on the State program before it is submitted to the Administrator.

(c) DISAPPROVAL

(1) IN GENERAL—If, in the judgment of the Administrator, a State program (or portion thereof, including the definition of a wellhead protection area), is not adequate to protect public water systems as required by this section, the Administrator shall disapprove such program (or portion thereof). A State program

developed pursuant to subsection (a) shall be deemed to be adequate unless the Administrator determines, within 9 months of the receipt of a State program, that such program (or portion thereof) is inadequate for the purpose of protecting public water systems as required by this section from contaminants that may have any adverse effect on the health of persons. If the Administrator determines that a proposed State program (or any portion thereof) is inadequate, the Administrator shall submit a written statement of the reasons for such determination to the Governor of the State.

(2) MODIFICATION AND RESUBMISSION – Within 6 months after receipt of the Administrator's written notice under paragraph (1) that any proposed State program (or portion thereof) is inadequate, the Governor or Governor's designed, shall modify the program based upon the recommendations of the Administrator and resubmit the modified program to the Administrator.

(d) FEDERAL ASSISTANCE – After the date 3 years after the enactment of this section, no State shall receive funds authorized to be appropriated under this section except for the purpose of implementing the program and requirements of paragraphs (4) and (6) of subsection (a).

(e) DEFINITION OF WELLHEAD PROTECTION AREA – As used in this section, the term "wellhead protection area" means the surface and subsurface area surrounding a water well or wellfield, supplying a public water system, through which contaminants are reasonably likely to move toward and reach such water well or wellfield. The extent of a wellhead protection area, within a State, necessary to provide protection from contaminants which may have any adverse effect on the health of persons is to be determined by the State in the program submitted under subsection (a). Not later than one year after the enactment of the Safe Drinking Water Act Amendments of 1986, the Administrator shall issue technical guidance which States may use in making such determinations. Such guidance may reflect such factors as the radius of influence around a well or wellfield, the depth of drawdown of the water table by such well or wellfield at any given point, the time or rate of travel of various contaminants in various hydrologic conditions, distance from the well or wellfield, or other factors affecting the likelihood of contaminants reaching the well or wellfield, taking into account available engineering pump tests or comparable data, field reconnaissance, topographic information, and the geology of the formation in which the well or wellfield is located.

(f) PROHIBITIONS –

(1) ACTIVITIES UNDER OTHER LAWS – No funds authorized to be appropriated under this section may be used to support activities authorized by the Federal Water Pollution Control Act, the Solid Waste Disposal Act, the Comprehensive Environmental Response, Compensation, and Liability Act of 1980, or other sections of this Act.

(2) INDIVIDUAL SOURCES – No funds authorized to be appropriated under this section may be used to bring individual sources of contamination into compliance.

(g) IMPLEMENTATION – Each State shall make every reasonable effort to implement the State wellhead area protection program under this section within 2

years of submitting the program to the Administrator. Each State shall submit to the Administrator a biennial status report describing the State's progress in implementing the program. Such report shall include amendments to the State program for water wells sited during the biennial period.

(h) FEDERAL AGENCIES—Each department, agency, and instrumentality of the executive, legislative, and judicial branches of the Federal Government having jurisdiction over any potential source of contaminants identified by a State program pursuant to the provisions of subsection (a)(3) shall be subject to and comply with all requirements of the State program developed according to subsection (a)(4) applicable to such potential source of contaminants, both substantive and procedural, in the same manner, and to the same extent, as any other person is subject to such requirements, including payment of reasonable charges and fees. The President may exempt any potential source under the jurisdiction of any department, agency, or instrumentality in the executive branch if the President determines it to be in the paramount interest of the United States to do so. No such exemption shall be granted due to the lack of an appropriation unless the President shall have specifically requested such appropriation as part of the budgetary process and the Congress shall have failed to make available such requested appropriations.

(i) ADDITIONAL REQUIREMENT—

(1) IN GENERAL—In addition to the provisions of subsection (a) of this section, States in which there are more than 2,500 active wells at which annular injection is used as of January 1, 1986, shall include in their State program a certification that a State program exists and is being adequately enforced that provides protection from contaminants which may have any adverse effect on the health of persons and which are associated with the annular injection or surface disposal of brines associated with oil and gas production.

(2) DEFINITION—For purposes of this subsection, the term "annular injection" means the reinjection of brines associated with the production of oil or gas between the production and surface casings of a conventional oil or gas producing well.

(3) REVIEW—The Administrator shall conduct a review of each program certified under this section.

(4) DISAPPROVAL—If a State fails to include the certification required by this subsection or if in the judgment of the Administrator the State program certified under this subsection is not being adequately enforced, the Administrator shall disapprove the State program submitted under subsection (a) of this section.

(j) COORDINATION WITH OTHER LAWS—Nothing in this section shall authorize or require any department, agency, or other instrumentality of the Federal Government or State or local government to apportion, allocate or otherwise regulate the withdrawal or beneficial use of ground or surface waters, so as to abrogate or modify any existing rights to water established pursuant to State or Federal law, including interstate compacts.

(k) AUTHORIZATION OF APPROPRIATIONS—Unless the State program is disapproved under this section, the Administrator shall make grants to the

State for not less than 50 or more than 90 percent of the costs incurred by a State (as determined by the Administrator) in developing and implementing each State program under this section. For purposes of making such grants there is authorized to be appropriated not more than the following amounts:

Fiscal year:	**Amount**
1987	**$20,000,000**
1988	**20,000,000**
1989	**35,000,000**
1990	**35,000,000**
1991	**35,000,000**

Part D—Emergency Powers

EMERGENCY POWERS

Section 1431. (a) Notwithstanding any other provision of this title, the Administrator, upon receipt of information that a contaminant which is present in or is likely to enter a public water system **or an underground source of drinking water** may present an imminent and substantial endangerment to the health of persons, and that appropriate State and local authorities have not acted to protect the health of such persons, may take such actions as he may deem necessary in order to protect the health of such persons. To the extent he determines it to be practicable in light of such imminent endangerment, he shall consult with the State and local authorities in order to confirm the correctness of the information on which action proposed to be taken under this subsection is based and to ascertain the action which such authorities are or will be taking. The action which the Administrator may take may include (but shall not be limited to) (1) issuing such orders as may be necessary to protect the health of persons who are or may be users of such system (including travelers), **including orders requiring the provision of alternative water supplies by persons who caused or contributed to the endangerment,** and (2) commencing a civil action for appropriate relief, including a restraining order or permanent or temporary injunction.

(b) Any person who violates or fails or refuses to comply with any order issued by the Administrator under subsection (a)(1) may, in an action brought in the appropriate United States district court to enforce such order, be **subject to a civil penalty of not to exceed** $5,000 for each day in which such violation occurs or failure to comply continues.

Section 1432. TAMPERING WITH PUBLIC WATER SYSTEMS

(a) TAMPERING—Any person who tampers with a public water system shall be imprisoned for not more than 5 years, or fined in accordance with title 18 of the United States Code, or both.

(b) ATTEMPT OR THREAT—Any person who attempts to tamper, or makes a threat to tamper, with a public drinking water system be imprisoned for not more than 3 years, or fined in accordance with title 18 of the United States Code.

(c) CIVIL PENALTY—The Administrator may bring a civil action in the appropriate United States district court (as determined under the provisions of title 28 of the United States Code) against any person who tampers, attempts to tamper, or makes a threat to tamper with a public water system. The court may impose on such person a civil penalty of not more than $50,000 for such tampering or, not more than $20,000 for such attempt or threat.

(d) DEFINITION OF "TAMPER"—For purposes of this section, the term "tamper" means—

(1) to introduce a contaminant into a public water system with the intention of harming persons; or

(2) to otherwise interfere with the operation of a public water system with the intention of harming persons.

Part E—General Provisions

ASSURANCES OF AVAILABILITY OF ADEQUATE SUPPLIES OF CHEMICALS NECESSARY FOR TREATMENT OF WATER

Section 1441. (a) If any person who uses chlorine, activated carbon, lime, ammonia, soda ash, potassium permanganate, caustic soda, or other chemical or substance for the purpose of treating water in any public water system or in any public treatment works determines that the amount of such chemical or substance necessary to effectively treat such water is not reasonably available to him or will not be so available to him when required for the effective treatment of such water, such person may apply to the Administrator for a certification (hereinafter in this section referred to as a "certification of need") that the amount of such chemical or substance which such person requires to effectively treat such water is not reasonably available to him or will not be so available when required for the effective treatment of such water.

(b)(1) An application for a certification of need shall be in such form and submitted in such manner as the Administrator may require and shall (A) specify the persons the applicant determines are able to provide the chemical or substance with respect to which the application is submitted, (B) specify the persons from whom the applicant has sought such chemical or substance, and (C) contain such other information as the Administrator may require.

(2) Upon receipt of an application under this section, the Administrator shall (A) publish in the Federal Register a notice of the receipt of the application and a brief summary of it, (B) notify in writing each person whom the President or his delegate (after consultation with the Administrator) determines could be made subject to an order required to be issued upon the issuance of the certification of need applied for in such application, and (C) provide an opportunity for the submission of written comments on such application. The requirements of the preceding sentence of this paragraph shall not apply when the Administrator for good cause finds (and incorporates the finding with a brief statement of reasons therefor in the order issued) that waiver of such requirements is necessary in order to protect the public health.

(3) Within 30 days after—

(A) the date a notice is published under paragraph (2) in the Federal Register with respect to an application submitted under this section for the issuance of a certification of need, or

(B) the date on which such application is received if as authorized by the second sentence of such paragraph no notice is published with respect to such application,

the Administrator shall take action either to issue or deny the issuance of a certification of need.

(c)(1) If the Administrator finds that the amount of a chemical or substance necessary for an applicant under an application submitted under this section to effectively treat water in a public water system or in a public treatment works is not reasonably available to the applicant or will not be so available to him when required for the effective treatment of such water, the Administrator shall issue a certification of need. Not later than seven days following the issuance of such certification, the President or his delegate shall issue an order requiring the provision to such person of such amounts of such chemical or substance as the Administrator deems necessary in the certification of need issued for such person. Such order shall apply to such manufacturers, producers, processors, distributors, and repackagers of such chemical or substance as the President or his delegate deems necessary and appropriate, except that such order may not apply to any manufacturer, producer, or processor of such chemical or substance who manufactures, produces, or processes (as the case may be) such chemical or substance solely for its own use. Persons subject to an order issued under this section shall be given a reasonable opportunity to consult with the President or his delegate with respect to the implementation of the order.

(2) Orders which are to be issued under paragraph (1) to manufacturers, producers, and processors of a chemical or substance shall be equitably apportioned, as far as practicable, among all manufacturers, producers, and processors of such chemical or substance; and orders which are to be issued under paragraph (1) to distributors and repackagers of a chemical or substance shall be equitably apportioned, as far as practicable, among all distributions and repackagers of such chemical or substance. In apportioning orders issued under paragraph (1) to manufacturers, producers, processors, distributors, and repackagers of chlorine, the President or his delegate shall, in carrying out the requirements of the preceding sentence, consider—

(A) the geographical relationship and established commercial relationships between such manufacturers, producers, processors, distributors, and repackagers and the persons for whom the orders are issued;

(B) in the case of orders to be issued to producers of chlorine, the (i) amount of chlorine historically supplied by each such producer to treat water in public water systems and public treatment works, and (ii) share of each such producer of the total annual production of chlorine in the United States; and

(C) such other factors as the President or his delegate may determine are relevant to the apportionment of orders in accordance with the requirements of the preceding sentence.

(3) Subject to subsection (f), any person for whom a certification of need has

been issued under this subsection may upon the expiration of the order issued under paragraph (1) upon such certification apply under this section for additional certifications.

(d) There shall be available as a defense to any action brought for breach of contract in a Federal or State court arising out of delay or failure to provide, sell, or offer for sale or exchange a chemical or substance subject to an order issued pursuant to subsection (c)(1), that such delay or failure was caused solely by compliance with such order.

(e)(1) Whoever knowingly fails to comply with any order issued pursuant to subsection (c)(1) shall be fined not more than $5,000 for each such failure to comply.

(2) Whoever fails to comply with any order issued pursuant to subsection (c)(1) shall be subject to a civil penalty of not more than $2,500 for each such failure to comply.

(3) Whenever the Administrator or the President or his delegate has reason to believe that any person is violating or will violate any order issued pursuant to subsection (c)(1), he may petition a United States district court to issue a temporary restraining order or preliminary or permanent injunction (including a mandatory injunction) to enforce the provisions of such order.

(f) No certification of need or order issued under this section may remain in effect **for more than one year.**

RESEARCH, TECHNICAL ASSISTANCE, INFORMATION, AND TRAINING OF PERSONNEL

Section 1442. (a)(1) The Administrator may conduct research, studies, and demonstrations relating to the causes, diagnosis, treatment, control, and prevention of physical and mental diseases and other impairments of man resulting directly or indirectly from contaminants in water, or to the provision of a dependably safe supply of drinking water, including—

(A) improved methods (i) to identify and measure the existence of contaminants in drinking water (including methods which may be used by State and local health and water officials), and (ii) to identify the source of such contaminants;

(B) improved methods to identify and measure the health effects of contaminants in drinking water;

(C) new methods of treating raw water to prepare it for drinking, so as to improve the efficiency of water treatment and to remove contaminants from water;

(D) improved methods for providing a dependably safe supply of drinking water, including improvements in water purification and distribution, and methods of assessing the health related hazards of drinking water; and

(E) improve methods of protecting underground water sources of public water systems from contamination.

(2)(A) The Administrator shall, to the maximum extent feasible, provide technical assistance to the States and municipalities in the establishment and administration of public water system supervision programs (as defined in section 1443(c)(1)).

(B) The Administrator is authorized to provide technical assistance and to make grants to States, or publicly owned water systems to assist in responding to and alleviating any emergency situation affecting public water systems (including sources of water for such systems) which the Administrator determines to present substantial danger to the public health. Grants provided under this subparagraph shall be used only to support those actions which (i) are necessary for preventing, limiting or mitigating danger to the public health in such emergency situation and (ii) would not, in the judgment of the Administrator, be taken without such emergency assistance. The Administrator may carry out the program authorized under this subparagraph as part of, and in accordance with the terms and conditions of, any other program of assistance for environmental emergencies which the Administrator is authorized to carry out under any other provision of law. No limitation on appropriations for any such other program shall apply to amounts appropriated under this subparagraph.

(3)(A) The Administrator shall conduct studies, and make periodic reports to Congress, on the costs of carrying out regulations prescribed under section 1412.

(B) Not later than eighteen months after the date of enactment of this subparagraph, the Administrator shall submit a report to Congress which identifies and analyzes—

(i) the anticipated costs of compliance with interim and revised national primary drinking water regulations and the anticipated costs to States and units of local governments in implementing such regulations;

(ii) alternative methods of (including alternative treatment techniques for) compliance with such regulations;

(iii) methods of paying the costs of compliance by public water systems with national primary drinking water regulations, including user charges, State or local taxes or subsidies, Federal grants (including planning or construction grants, or both), loans, and loan guarantees, and other methods of assisting in paying the costs of such compliance;

(iv) the advantages and disadvantages of each of the methods referred to in clauses (ii) and (iii);

(v) the sources of revenue presently available (and projected to be available) to public water systems to meet current and future expenses; and

(vi) the costs of drinking water paid by residential and industrial consumers in a sample of large, medium, and small public water systems and of individually owned wells, and the reasons for any differences in such costs.

The report required by this subparagraph shall identify and analyze the items required in clauses (i) through (v) separately with respect to public water systems serving small communities. The report required by this subparagraph shall include such recommendations as the Administrator deems appropriate.

(4) The Administrator shall conduct a survey and study of—

(A) disposal of waste (including residential waste) which may endanger underground water which supplies, or can reasonably be expected to supply, any public water systems, and

(B) means of control of such waste disposal.

Not later than one year after the date of enactment of this title, he shall transmit to the Congress the results of such survey and study, together with such recommendations as he deems appropriate.

(5) The Administrator shall carry out a study of methods of underground injection which do not result in the degradation of underground drinking water sources.

(6) The Administrator shall carry out a study of methods of preventing, detecting, and dealing with surface spills of contaminants which may degrade underground water sources for public water systems.

(7) The Administrator shall carry out a study of virus contamination of drinking water sources and means of control of such contamination.

(8) The Administrator shall carry out a study of the nature and extent of the impact on underground water which supplies or can reasonably be expected to supply public water systems of (A) abandoned injection or extraction wells; (B) intensive application of pesticides and fertilizers in underground water recharge areas; and (C) ponds, pools, lagoons, pits, or other surface disposal of contaminants in underground water recharge areas.

(9) The Administrator shall conduct a comprehensive study of public water supplies and drinking water sources to determine the nature, extent, sources of and means of control of contamination by chemicals or other substances suspected of being carcinogenic. Not later than six months after the date of enactment of this title, he shall transmit to the Congress the initial results of such study, together with such recommendations for further review and corrective action as he deems appropriate.

(10) The Administrator shall carry out a study of the reaction of chlorine and humic acids and the effects of the contaminants which result from such reaction on public health and on the safety of drinking water, including any carcinogenic effect.

(11) The Administrator shall carry out a study of polychlorinated biphenyl contamination of actual or potential sources of drinking water, contamination of such sources by other substances known or suspected to be harmful to public health, the effects of such contamination, and means of removing, treating, or otherwise controlling such contamination. To assist in carrying out this paragraph, the Administrator is authorized to make grants to public agencies and private nonprofit institutions.

(b) In carrying out this title, the Administrator is authorized to—

(1) collect and make available information pertaining to research, investigations, and demonstrations with respect to providing a dependably safe supply of

drinking water together with appropriate recommendations in connection therewith;

(2) make available research facilities of the Agency to appropriate public authorities, institutions, and individuals engaged in studies and research relating to the purposes of this title;

(3) make grants to and enter into contracts with, any public agency, educational institution, and any other organization, in accordance with procedures prescribed by the Administrator, under which he may pay all or part of the costs (as may be determined by the Administrator) of any project or activity which is designed—

(A) to develop, expand, or carry out a program (which may combine training education and employment) for training persons for occupations involving the public health aspects of providing safe drinking water;

(B) to train inspectors and supervisory personnel to train or supervise persons in occupations involving the public health aspects of providing safe drinking water; or

(C) to develop and expand the capability of programs of State and municipalities to carry out the purposes of this title (other than by carrying out State programs of public water system supervision or underground water source protection (as defined in section 1443(c))).

(c) Not later than eighteen months after the date of enactment of this subsection, the Administrator shall submit a report to Congress on the present and projected future availability of an adequate and dependable supply of safe drinking water to meet present and projected future need. Such report shall include an analysis of the future demand for drinking water and other competing uses of water, the availability and use of methods to conserve water or reduce demand, the adequacy of present measures to assure adequate and dependable supplies of safe drinking water, and the problems (financial, legal, or other) which need to be resolved in order to assure the availability of such supplies for the future. Existing information and data compiled by the National Water Commission and others shall be utilized to the extent possible.

(d) The Administrator shall—

(1) provide training for, and make grants for training (including postgraduate training) of (A) personnel of State agencies which have primary enforcement responsibility and of agencies of units of local government to which enforcement responsibilities have been delegated by the State, and (B) personnel who manage or operate public water systems, and

(2) make grants for postgraduate training of individuals (including grants to educational institutions for traineeships) for purposes or qualifying such individuals to work as personnel referred to in paragraph (1).

Reasonable fees may be charged for training provided under paragraph (1)(B) to persons other than personnel of State or local agencies but such training shall be provided to personnel of State or local agencies without charge.

(e) Repealed.

(f) There are authorized to be appropriated to carry out the provisions is section, other than subsection (a)(2)(B) and provisions relating to research,

$15,000,000 for the fiscal year ending June 30, 1975; $25,000,000 for the fiscal year ending June 30, 1976; $35,000,000 for the fiscal year ending June 30, 1977; $17,000,000 for each of the fiscal years 1978 and 1979; $21,405,000 for the fiscal year ending September 30, 1980; $30,000,000 for the fiscal year ending September 30, 1981; $35,000,000 for the fiscal year ending September 30, 1982. There are authorized to be appropriated to carry out subsection (a)(2)(B) $8,000,000 for each of the fiscal years 1978 through 1982. **There are authorized to be appropriated to carry out subsection (a)(2)(B) not more than the following amounts:**

Fiscal year:	**Amount**
1987	**$7,650,000**
1988	**7,650,000**
1989	**8,050,000**
1990	**8,050,000**
1991	**8,050,000**

There are authorized to be appropriated to carry out the provisions of this section (other than subsection (g), subsection (a)(2)(B), and provisions relating to research), not more than the following amounts:

Fiscal year:	**Amount**
1987	**$35,600,000**
1988	**35,600,000**
1989	**38,020,000**
1990	**38,020,000**
1991	**38,020,000**

(g) The Administrator is authorized to provide technical assistance to small public water systems to enable such systems to achieve and maintain compliance with national drinking water regulations. Such assistance may include "circuit-rider" programs, training, and preliminary engineering studies. There are authorized to be appropriated to carry out this subsection $10,000,000 for each of the fiscal years 1987 through 1991. Not less than the greater of—

(1) 3 percent of the amounts appropriated under this subsection, or

(2) $280,000 shall be utilized for technical assistance to public water systems owned or operated by Indian tribes.

GRANTS FOR STATE PROGRAMS

Section 1443. (a)(1) From allotments made pursuant to paragraph (4), the Administrator may make grants to States to carry out public water system supervision programs.

(2) No grant may be made under paragraph (1) unless an application therefor has been submitted to the Administrator in such form and manner as he may require. The Administrator may not approve an application of a State for its first grant under paragraph (1) unless he determines that the State—

(A) has established or will establish within one year from the date of such grant a public water system supervision program, and

(B) will, within that one year, assume primary enforcement responsibility for public water system within the State.

No grant may be made to a State under paragraph (1) for any period beginning more than one year after the date of the State's first grant unless the State has assumed and maintained primary enforcement responsibility for public water systems within the State. **The prohibitions contained in the preceding two sentences shall not apply to such grants when made to Indian Tribes.**

(3) A grant under paragraph (1) shall be made to cover not more than 75 per centum of the grant recipient's costs (as determined under regulations of the Administrator) in carrying out, during the one-year period beginning on the date the grant is made, a public water system supervision program.

(4) In each fiscal year the Administrator shall, in accordance with regulations, allot the sums appropriated for such year under paragraph (5) among the States on the basis of population, geographical area, number of public water systems, and other relevant factors. No State shall receive less than 1 per centum of the annual appropriation for grants under paragraph (1): *Provided,* That the Administrator may by regulation, reduce such percentage in accordance with the criteria specified in this paragraph: *And provided further,* That such percentage shall not apply to grants allotted to Guam, American Samoa, or the Virgin Islands.

(5) The prohibition contained in the last sentence of paragraph (2) may be waived by the Administrator with respect to a grant to a State through fiscal year 1979 but such prohibition may only be waived if, in the judgment of the Administrator—

(A) the State is making diligent effort to assume and maintain primary enforcement responsibility for public water systems within the State;

(B) the State has made significant progress toward assuming and maintaining such primary enforcement responsibility; and

(C) there is reason to believe the State will assume such primary enforcement responsibility by October 1, 1979.

The amount of any grant awarded for the fiscal years 1978 and 1979 pursuant to a waiver under this paragraph may not exceed 75 per centum of the allotment which the State would have received for such fiscal year if it had assumed and maintained such primary enforcement responsibility. The remaining 25 per centum of the amount allotted to such State for such fiscal year shall be retained by the Administrator, and the Administrator may award such amount to such State at such time as the State assumes such responsibility before the beginning of fiscal year 1980. At the beginning of each fiscal years 1979 and 1980 the amounts retained by the Administrator for any preceding fiscal year and not awarded by the beginning fiscal year 1979 or 1980 to the States to which such amounts were originally allotted may be removed from the original allotment and reallotted for fiscal year 1979 or 1980 (as the case may be) to States which have assumed primary enforcement responsibility by the beginning of such fiscal year.

(6) The Administrator shall notify the State of the approval or disapproval of any application for a grant under this section—

(A) within ninety days after receipt of such application, or

(B) not later than the first day of the fiscal year for which the grant application is made,

whichever is later.

(7) For the purposes of making grants under paragraph (1) there are authorized to be appropriated $15,000,000 for the fiscal year ending June 30, 1976, $25,000,000 for the fiscal year ending June 30, 1977, $35,000,000 for fiscal year 1978, $45,000,000 for fiscal year 1979, $29,450,000 for the fiscal year ending September 30, 1980, $32,000,000 for the fiscal year ending September 30, 1981, and $34,000,000 for the fiscal year ending September 30, 1982. **For the purposes of making grants under paragraph (1) there are authorized to be appropriated not more than the following amounts:**

Fiscal year:	**Amount**
1987	**$37,200,000**
1988	**37,200,000**
1989	**40,150,000**
1990	**40,150,000**
1991	**40,150,000**

(b)(1) From allotments made pursuant to paragraph (4), the Administrator may make grants to States to carry out underground water source protection programs.

(2) No grant may be made under paragraph (1) unless an application therefor has been submitted to the Administrator in such form and manner as he may require. No grant may be made to any State under paragraph (1) unless the State has assumed primary enforcement responsibility within two years after the date the Administrator promulgates regulations for State underground injection control programs under section 1421. **The prohibition contained in the preceding sentence shall not apply to such grants when made to Indian Tribes.**

(3) A grant under paragraph (1) shall be made to cover not more than 75 per centum of the grant recipient's costs (as determined under regulations of the Administrator) in carrying out, during the one-year period beginning on the date the grant is made, an underground water source protection program.

(4) In each fiscal year the Administrator shall, in accordance with regulations, allot the sums appropriated for such year under paragraph (5) among the States on the basis of population, geographical area, and other relevant factors.

(5) For purposes of making grants under paragraph (1) there are authorized to be appropriated $5,000,000 for the fiscal year ending June 30, 1976, $7,500,000 for the fiscal year ending June 30, 1977, $10,000,000 for each of the fiscal years 1978 and 1979, $7,795,000 for the fiscal year ending September 30, 1980, $18,000,000 for the fiscal year ending September 30, 1981, and $21,000,000 for the fiscal year ending September 30, 1982. **For the purpose of making grants under paragraph (1) there are authorized to be appropriated not more than the following amounts:**

Fiscal year:	Amount
1987	$19,700,000
1988	19,700,000
1989	20,850,000
1990	20,850,000
1991	20,850,000

(c) For purposes of this section:

(1) the term "public water system supervision program" means a program for the adoption and enforcement of drinking water regulations (with such variances and exemptions from such regulations under conditions and in a manner which is not less stringent than the conditions under, and the manner in, which variances and exemptions may be granted under section 1415 and 1416) which are no less stringent than the national primary drinking water regulations under section 1412, and for keeping records and making reports required by section 1413(a)(3).

(2) The term "underground water source protection program" means a program for the adoption and enforcement of a program which meets the requirements of regulations under section 1421 and for keeping records and making reports required by section 1422(b)(1)(A)(ii). Such term includes, where applicable, a program which meets the requirements of section 1425.

SPECIAL STUDY AND DEMONSTRATION PROJECT GRANTS; GUARANTEED LOANS

Section 1444. (a) The Administrator may make grants to any person for the purposes of—

(1) assisting in the development and demonstration (including construction) of any project which will demonstrate a new or improved method, approach, or technology, for providing a dependably safe supply of drinking water to the public; and

(2) assisting in the development and demonstration (including construction) of any project which will investigate and demonstrate health implications involved in the reclamation, recycling, and reuse of waste waters for drinking and the processes and methods for the preparation of safe and acceptable drinking water.

(b) Grants made by the Administrator under this section shall be subject to the following limitations:

(1) Grants under this section shall not exceed $66^2/_3$ per centum of the total cost of construction of any facility, and 75 per centum of any other costs, as determined by the Administrator.

(2) Grants under this section shall not be made for any project involving the construction or modification of any facilities for any public water system in a State unless such project has been approved by the State agency charged with the responsibility for safety of drinking water (or if there is no such agency in a State, by the State health authority).

(3) Grants under this section shall not be made for any project unless the Administrator determines, after consulting the National Drinking Water Advisory Council, that such project will serve a useful purpose relating to the development and demonstration of new or improved techniques, methods, or technologies for the provision of safe water to the public for drinking.

(4) Priority for grants under this section shall be given where there are known or potential public health hazards which require advanced technology for the removal of particles which are too small to be removed by ordinary treatment technology.

(c) For the purposes of making grants under subsections (a) and (b) of this section there are authorized to be appropriated $7,500,000 for the fiscal year ending June 30, 1975; and $7,500,000 for the fiscal year ending June 30, 1976; and $10,000,000 for the fiscal year ending June 30, 1977.

(d) The Administrator during the fiscal years ending June 30, 1975, and June 30, 1976, shall carry out a program of guaranteeing loans made by private lenders to small public water systems for the purpose of enabling such systems to meet national primary drinking water regulations prescribed under section 1412. No such guarantee may be made with respect to a system unless (1) such system cannot reasonably obtain financial assistance necessary to comply with regulations from any other source, and (2) the Administrator determines that any facilities constructed with a loan guaranteed under this subsection is not likely to be made obsolete by subsequent changes in primary regulations. The aggregate amount of indebtedness guaranteed with respect to any system may not exceed $50,000. The aggregate amount of indebtedness guaranteed under this subsection may not exceed $50,000,000. The Administrator shall prescribe regulations to carry out this subsection.

RECORDS AND INSPECTIONS

Section 1445. (a)(1) Every person who is a supplier of water, who is or may be otherwise subject to a primary drinking water regulation prescribed under section 1412 or to an applicable underground injection control program (as defined in section 1422(c)), who is or may be subject to the permit requirement of section 1424 or to an order issued under section 1441, or who is a grantee, shall establish and maintain such records, make such reports, conduct such monitoring, and provide such information as the Administrator may reasonably require by regulation to assist him in establishing regulations under this title, in determining whether such person has acted or is acting in compliance with this title, in administering any program of financial assistance under this title, in evaluating the health risks of unregulated contaminants, or in advising the public of such risks. **In requiring a public water system to monitor under this subsection, the Administrator may take into consideration the system size and the contaminants likely to be found in the system's drinking water.**

(2) Not later than 18 months after enactment of the Safe Drinking Water Act Amendments of 1986, the Administrator shall promulgate regulations requiring every public water system to conduct a monitoring program for unregulated contaminants. The regulations shall require monitoring of drinking water sup-

plied by the system and shall vary the frequency and schedule of monitoring requirements for systems based on the number of persons served by the system, the source of supply, and the contaminants likely to be found. Each system shall be required to monitor at least once every 5 years after the effective date of the Administrator's regulations unless the Administrator requires more frequent monitoring.

(3) Regulations under paragraph (2) shall list unregulated contaminants for which systems may be required to monitor, and shall include criteria by which the primary enforcement authority in each State could show cause for addition or deletion of contaminants from the designated list. The primary State enforcement authority may delete contaminants for an individual system, in accordance with these criteria, after obtaining approval of assessment of the contaminants potentially to be found in the system. The Administrator shall approve or disapprove such an assessment submitted by a State within 60 days. A State may add contaminants, in accordance with these criteria, without making an assessment, but in no event shall such additions increase Federal expenditures authorized by this section.

(4) Public water systems conducting monitoring of unregulated contaminants pursuant to this section shall provide the results of such monitoring to the primary enforcement authority.

(5) Notification of the availability of the results of the monitoring programs required under paragraph (2), and notification of the availability of the results of the monitoring program referred to in paragraph (6), shall be given to the persons served by the system and the Administrator.

(6) The Administrator may waive the monitoring requirement under paragraph (2) for a system which has conducted a monitoring program after January 1, 1983, if the Administrator determines the program to have been consistent with the regulations promulgated under this section.

(7) Any system supplying less than 150 service connections shall be treated as complying with this subsection if such system provides water samples or the opportunity for sampling according to rules established by the Administrator.

(8) There are authorized to be appropriated $30,000,000 in the fiscal year ending September 30, 1987 to remain available until expended to carry out the provisions of this subsection.

(b)(1) Except as provided in paragraph (2), the Administrator, or representatives of the Administrator duly designated by him, upon presenting appropriate credentials and a written notice to any supplier of water or other person subject to (A) a national primary drinking water regulation prescribed under section 1412, (B) and applicable underground injection control program, or (C) any requirement to monitor an unregulated contaminant pursuant to subsection (a), or person in charge of any of the property of such supplier or other person referred to in clause (A), (B), or (C), is authorized to enter any establishment, facility, or other property of such supplier or other person in order to determine whether such supplier or other person has acted or is acting in compliance with this title, including for this purpose, inspection, at reasonable times, of records, files, papers, processes, controls, and facilities, or in order to test any feature of a public water system, including its raw water source. The Administrator or the Comptroller General (or any representative designated by either) shall have access for the purpose of audit and examination to any records, reports, or information of a grantee which are required to be maintained under subsection (a) or which are pertinent to any financial assistance under this title.

(2) No entry may be made under the first sentence of paragraph (1) in an establishment, facility, or other property of a supplier of water or other person subject to a national primary drinking water regulation if the establishment, facility, or other property is located in a State which has primary enforcement responsibility for public water systems unless, before written notice of such entry is made, the Administrator (or his representative) notifies the State agency charged with responsibility for safe drinking water of the reasons for such entry. The Administrator shall, upon a showing by the State agency that such an entry will be detrimental to the administration of the State's program of primary enforcement responsibility, take such showing into consideration in determining whether to make such entry. No State agency which receives notice under this paragraph of an entry proposed to be made under paragraph (1) may use the information contained in the notice to inform the person whose property is proposed to be entered of the proposed entry; and if a State agency so uses such information, notice to the agency under this paragraph is not required until such time as the Administrator determines the agency has provided him satisfactory assurances that it will no longer so use information contained in a notice under this paragraph.

(c) Whoever fails or refuses to comply with any requirement of subsection (a) or to allow the Administrator, the Comptroller General, or representatives of either, to enter and conduct any audit of inspection authorized by subsection (b) **shall be subject to a civil penalty of not to exceed $25,000.**

(d)(1) Subject to paragraph (2), upon a showing satisfactory to the Administrator by any person that any information required under this section from such person, if made public, would divulge trade secrets or secret processes of such person, the Administrator shall consider such information confidential in accordance with the purposes of section 1905 of title 18 of the United States Code. If the applicant fails to make a showing satisfactory to the Administrator, the Administrator shall give such applicant thirty days' notice before releasing the information to which the application relates (unless the public health or safety requires an earlier release of such information).

(2) Any information required under this section (A) may be disclosed to other officers, employees, or authorized representatives of the United States concerned with carrying out this title or to committees of the Congress, or when relevant in any proceeding under this title, and (B) shall be disclosed to the extent it deals with the level of contaminants in drinking water. For purposes of this subsection the term "information required under this section" means any papers, books, documents, or information, or any particular part thereof, reported to or otherwise obtained by the Administrator under this section.

(e) For purposes of this section, (1) the term "grantee" means any person who applies for or receives financial assistance, by grant, contract, or loan guarantee under this title, and (2) the term "person" includes a Federal agency.

NATIONAL DRINKING WATER ADVISORY COUNCIL

Section 1446. (a) There is established a National Drinking Water Advisory Council which shall consist of fifteen members appointed by the Administrator after consultation with the Secretary. Five members shall be appointed from the general public; five members shall be appointed from appropriate State and local agencies concerned with water hygiene and public water supply; and five members shall be appointed from representatives of private organizations or groups demonstrating an active interest in the field of water hygiene and public water supply. Each member of the Council shall hold office for a term of three years, except that—

(1) any member appointed to fill a vacancy occurring prior to the expiration of the term for which his predecessor was appointed shall be appointed for the remainder of such term; and

(2) the terms of the members first taking office shall expire as follows: Five shall expire three years after the date of enactment of this title, five shall expire two years after such date, and five shall expire one year after such date, as designated by the Administrator at the time of appointment.

The members of the Council shall be eligible for reappointment.

(b) The Council shall advise, consult with, and make recommendations to, the Administrator on matters relating to activities, functions, and policies of the Agency under this title.

(c) Members of the Council appointed under this section shall, while attending meetings or conferences of the Council or otherwise engaged in business of the Council, receive compensation and allowances at the rate to be fixed by the Administrator, but not exceeding the daily equivalent of the annual rate of basic pay in effect for grade GS-18 of the General Schedule for each day (including travel time) during which they are engaged in the actual performance of duties vested in the Council. While away from their homes or regular places of business in the performance of services for the Council, members of the Council shall be allowed travel expenses, including per diem in lieu of subsistence, in the same manner as persons employed intermittently in the Government service are allowed expenses under section 5703(b) or title 5 of the United States Code.

(d) Section 14(a) of the Federal Advisory Committee Act (relating to termination) shall not apply to the Council.

FEDERAL AGENCIES

Section 1447. (a) Each Federal agency (1) having jurisdiction over any federally owned or maintained public water system or (2) engaged in any activity resulting, or which may result in, underground injection which endangers drinking water (within the meaning of section 1421(d)(2)) shall be subject to, and comply with, all Federal, State, and local requirements, administrative authorities, and process and sanctions respecting the provision of safe drinking water and respecting any underground injection program in the same manner, and to the same extent, as any nongovernmental entity. The preceding sentence shall apply (A) to any requirement whether substantive or procedural (including any recordkeeping or reporting requirement, any requirement respecting permits, and any other requirement whatsoever), (B) to the exercise of any Federal, State, or local administrative authority, and (C) to any process or sanction, whether enforced in Federal, State, or local courts or in any other manner. This subsection shall apply, notwithstanding any immunity of such agencies, under any law or rule of law. No officer, agent, or employee of the United States shall be personally liable for any civil penalty under this title with respect to any act or omission within the scope of his official duties.

(b) The Administrator shall waive compliance with subsection (a) upon request of the Secretary of Defense and upon a determination by the President that the requested waiver is necessary in the interest of national security. The Administrator shall maintain a written record of the basis upon which such waiver was granted and make such record available for in camera examination when relevant in a judicial proceeding under this title. Upon the issuance of such a waiver, the Administrator shall publish in the Federal Register a notice that the waiver was granted for national security purposes, unless, upon the request of the Secretary of Defense, the Administrator determines to omit such publication because the publication itself would be contrary to the interests of national security, in which event the Administrator shall submit notice to the Armed Services Committee of the Senate and House of Representatives.

(c)(1) Nothing in the Safe Drinking Water Amendments of 1977 shall be construed to alter or affect the status of American Indian lands or water rights nor to waive any sovereignty over Indian lands guaranteed by treaty or statute.

(2) For the purposes of this Act, the term "Federal agency" shall be construed to refer to or include any American Indian tribe, nor to the Secretary of the Interior in his capacity as trustee of Indian lands.

JUDICIAL REVIEW

Section 1448. (a) A petition for review of—

(1) actions pertaining to the establishment of national primary drinking water regulations (including maximum contaminant level goals) may be filed only in the United States Court of Appeals for the District of Columbia circuit; and

(2) any other action of the Administrator under this Act may be filed in the circuit in which the petitioner resides or transacts business which is directly affected by the action.

Any such petition shall be filed within the 45-day period beginning on the date of the promulgation of the regulation or issuance of the order with respect to which review is sought or on the date of the determination with respect to which review is sought, and may be filed after the expiration of such 45-day period if the petition is based solely on grounds arising after the expiration of such period. Action of the Administrator with respect to which review could have been obtained under this subsection shall not be subject to judicial review in any civil or criminal proceeding for enforcement or in any civil action to enjoin enforcement.

(b) The United States district courts shall have jurisdiction of actions brought to review (1) the granting of, or the refusing to grant, a variance or exemption under section 1415 or 1416 or (2) the requirements of any schedule prescribed for a variance or exemption under such section or the failure to prescribe such a schedule. Such an action may only be brought upon a petition for review filed with the court within the 45-day period beginning on the date the action sought to be reviewed is taken or, in the case of a petition to review the refusal to grant a variance or exemption or the failure to prescribe a schedule, within the 45-day period beginning on the date action is required to be taken on the variance, exemption, or schedule, as the case may be. A petition for such review may be filed after the expiration of such period. Action with respect to which review could have been obtained under this subsection shall not be subject to judicial review in any civil or criminal proceeding for enforcement or in any civil action to enjoin enforcement.

(c) In any judicial proceeding in which review is sought of a determination under this title required to be made on the record after notice and opportunity for hearing, if any party applies to the court for leave to adduce additional evidence and shows to the satisfaction of the court that such additional evidence is material and that there were reasonable grounds for the failure to adduce such evidence in the proceeding before the Administrator, the court may order such additional evidence (and evidence in rebuttal thereof) to be taken before the Administrator, in such manner and upon such terms and conditions as the court may deem proper. The Administrator may modify his findings as to the facts, or make new findings, by reason of the additional evidence so taken, and he shall file such modified or new findings, and his recommendation, if any, for the modification or setting aside of his original determination, with the return of such additional evidence.

CITIZEN'S CIVIL ACTION

Section 1449. (a) Except as provided in subsection (b) of this section, any person may commence a civil action on his own behalf—

(1) against any person (including (A) the United States, and (B) any other governmental instrumentality or agency to the extent permitted by the eleventh amendment to the Constitution) who is alleged to be in violation of any requirement prescribed by or under this title, or

(2) against the Administrator where there is alleged a failure of the Administrator to perform any act or duty under this title which is not discretionary with the Administrator.

No action may be brought under paragraph (1) against a public water system for a violation of a requirement prescribed by or under this title which occurred within the 27-month period beginning on the first day of the month in which this title is enacted. The United States district courts shall have jurisdiction, without regard to the amount of controversy or the citizenship of the parties, to enforce in an action brought under this subsection any requirement prescribed by or under this title or to order the Administrator to perform an act, or duty described in paragraph (2), as the case may be.

(b) No civil action may be commenced—

(1) under subsection (a)(1) of this section respecting violation or a requirement prescribed by or under this title—

(A) prior to sixty days after the plaintiff has given notice of such violation (i) to the Administrator, (ii) to any alleged violator of such requirement and (iii) to the State in which the violation occurs, or

(B) if the Administrator, the Attorney General, or the State has commenced and is diligently prosecuting a civil action in a court of the United States to require compliance with such requirement, but in any such action in a court of the United States any person may intervene as a matter of right; or

(2) under subsection (a)(2) of this section prior to sixty days after the plaintiff has given notice of such action to the Administrator.

Notice required by this subsection shall be given in such manner as the Administrator shall prescribe by regulation. No person may commence a civil action under subsection (a) to require a State to prescribe a schedule under section 1415 or 1416 for a variance or exemption, unless such person shows to the satisfaction of the court that the State has in a substantial number of cases failed to prescribe such schedules.

(c) In any action under this section, the Administrator or the Attorney General, if not a party, may intervene as a matter of right.

(d) The court, in issuing any final order in any action brought under subsection (a) of this section, may award costs of litigation (including reasonable attorney and expert witness fees) to any party whenever the court determines such an award is appropriate. The court may, if a temporary restraining order or preliminary injunction is sought, require the filing of a bond or equivalent security in accordance with the Federal Rules of Civil Procedure.

(e) Nothing in this section shall restrict any right which any person (or class of persons) may have under any statute or common law to seek enforcement of any requirement prescribed by or under this title or to seek any other relief. Nothing in this section or in any other law of the United States shall be construed to prohibit, exclude, or restrict any State or local government from—

(1) bringing any action or obtaining any remedy or sanction in any State or local court, or

(2) bringing any administrative action or obtaining any administrative remedy or sanction, against any agency of the United States under State or local law to enforce any requirement respecting the provision of safe drinking water or respecting any underground injection control program. Nothing in this section shall be construed to authorize judicial review of regulations or orders of the Administrator under this title, except as provided in section 1448. For provisions providing for application of certain requirements to such agencies in the same manner as to non-governmental entities, see section 1447.

GENERAL PROVISIONS

Section 1450. (a)(1) The Administrator is authorized to prescribe such regulations as are necessary or appropriate to carry out his functions under this title.

(2) The Administrator may delegate any of his functions under this title (other than prescribing regulations) to any officer or employee of the Agency.

(b) The Administrator, with the consent of the head of any other agency of the United States, may utilize such officers and employees of such agency as he deems necessary to assist him in carrying out the purposes of this title.

(c) Upon the request of a State or interstate agency, the Administrator may assign personnel of the Agency to such State or interstate agency for the purposes of carrying out the provisions of this title.

(d)(1) The Administrator may make payments of grants under this title (after necessary adjustment on account of previously made underpayments or overpayments) in advance or by way of reimbursement, and in such installments and on such conditions as he may determine.

(2) Financial assistance may be made available in the form of grants only to individuals and nonprofit agencies or institutions. For purposes of this paragraph, the term "nonprofit agency or institution" means an agency or institution no part of the net earnings of which inure, or may lawfully inure, to the benefit of any private shareholder or individual.

(e) The Administrator shall take such action as may be necessary to assure compliance with provisions of the Act of March 3, 1931 (known as the Davis-Bacon Act; 40 U.S.C. 276a-276a(5)). The Secretary of Labor shall have, with respect to the labor standards specified in this subsection, the authority and functions set forth in Reorganization Plan Numbered 14 of 1950 (15 F.R. 3176; 64 Stat. 1267) and section 2 of the Act of June 13, 1934 (40 U.S.C. 276c).

(f) The Administrator shall request the Attorney General to appear and represent him in any civil action instituted under this title to which the Administrator is a party. Unless, within a reasonable time, the Attorney General notifies the Administrator that he will appear in such action, attorneys appointed by the Administrator shall appear and represent him.

(g) The provisions of this title shall not be construed as affecting any authority of the Administrator under part G of title III of this Act.

(h) Not later than April 1 of each year, the Administrator shall submit to the Committee on Commerce of the Senate and the Committee on Interstate and Foreign Commerce of the House of Representatives a report respecting the activities of the Agency under this title and containing such recommendations for legislation as he considers necessary. The report of the Administrator under this subsection which is due not later than April 1, 1975, and each subsequent report of the Administrator under this subsection shall include a statement on the actual and anticipated cost to public water systems in each State of compliance with the requirements of this title. The Office of Management and Budget may review any report required by this subsection before its submission to such committees of Congress, but the Office may not revise any such report, require any revision in any such report, or delay its submission beyond the day prescribed for its submission, and may submit to such committees of Congress its comments respecting any such report.

(i)(1) No employer may discharge any employee or otherwise discriminate against any employee with respect to his compensation, terms, conditions, or privileges of employment because the employee (or any person acting pursuant to a request of the employee) has—

(A) commenced, caused to be commenced, or is about to commence or cause to be commenced a proceeding under this title or a proceeding for the administration or enforcement of drinking water regulations or underground injection control programs of a State.

(B) testified or is about to testify in any such proceeding, or

(C) assisted or participated or is about to assist or participate in any manner in such proceeding or in other action to carry out the purposes or this title.

(2)(A) Any employee who believes that he has been discharged or other wise discriminated against by any person in violation of paragraph (1) may, within 30 days after such violation occurs, file (or have any person file on his behalf) a complaint with the Secretary of Labor (hereinafter in this subsection referred to as the "Secretary") alleging such discharge or discrimination. Upon receipt of such a complaint, the Secretary shall notify the person named in the complaint of the filing of the complaint.

(B)(i) Upon receipt of a complaint filed under subparagraph (A), the Secretary shall conduct an investigation of the violation alleged in the complaint. Within 30 days of the receipt of such complaint, the Secretary shall complete such investigation and shall notify in writing the complainant (and any person acting in his behalf) and the person alleged to have committed such violation of the results of the investigation conducted pursuant to this subparagraph. Within 90 days of the receipt of such complaint the secretary shall, unless the proceeding on the complaint is terminated by the Secretary on the basis of a settlement entered into by the Secretary and the person alleged to have committed such violation, issue an order either providing the relief prescribed by clause (ii) or denying the complaint. An order of the Secretary shall be made on the record after notice and opportunity for agency hearing. The Secretary may not enter into a settlement terminating a proceeding on a complaint without the participation and consent of the complainant.

(ii) If in response to a complaint filed under subparagraph (A) the Secretary determines that a violation of paragraph (1) has occurred, the Secretary shall order (I) the person who committed such violation to take affirmative action to abate the violation, (II) such person to reinstate the complainant to his former position together with the compensation (including back pay), terms, conditions, and privileges of his employment, (III) compensatory damages, and (IV) where appropriate, exemplary damages. If such an order is issued, the Secretary, at the request of the complainant, shall assess against the person against whom the order is issued a sum equal to the aggregate amount of all costs and expenses (including attorneys' fees) reasonably incurred, as determined by the Secretary, by the complainant for, or in connection with, the bringing of the complaint upon which the order was issued.

(3)(A) Any person adversely affected or aggrieved by an order issued under paragraph (2) may obtain review of the order in the United States Court of Appeals for the circuit in which the violation, with respect to which the order was issued, allegedly occurred. The petition for review must be filed within sixty days from the issuance of the Secretary's order. Review shall conform to chapter 7 of title 5 of the United States Code. The commencement of proceedings under this subparagraph shall not, unless ordered by the court, operate as a stay of the Secretary's order.

(B) An order of the Secretary with respect to which review could have been obtained under subparagraph (A) shall not be subject to judicial review in any criminal or other civil proceeding.

(4) Whenever a person has failed to comply with an order issued under paragraph (2)(B), the Secretary shall file a civil action in the United States District Court for the district in which the violation was found to occur to enforce such order. In actions brought under this paragraph, the district court shall have jurisdiction to grant all appropriate relief including, but not limited to, injunctive relief, compensatory, and exemplary damages.

(5) Any nondiscretionary duty imposed by this section is enforceable in mandamus proceeding brought under section 1361 of title 28 of the United States Code.

(6) Paragraph (1) shall not apply with respect to any employee who, acting without direction from his employer (or the employer's agent), deliberately causes a violation of any requirement of this title.

Section 1451. Indian Tribes

(a) IN GENERAL—Subject to the provisions of subsection (b), the Administrator—

(1) is authorized to treat Indian Tribes as States under this title,

(2) may delegate to such Tribes primary enforcement responsibility for public water systems and for underground injection control, and

(3) may provide such Tribes grant and contract assistance to carry out functions provided by this title.

(b) EPA REGULATIONS—

(1) SPECIFIC PROVISIONS—The Administrator shall, within 18 months after the enactment of the Safe Drinking Water Act Amendments of 1986, promulgate final regulations specifying those provisions of this title for which it is appropriate to treat Indian Tribes as States. Such treatment shall be authorized only if:

(A) the Indian Tribe is recognized by the Secretary of the Interior and has a governing body carrying out substantial governmental duties and powers;

(B) the functions to be exercised by the Indian Tribe are within the area of the Tribal Government's jurisdiction; and

(C) the Indian Tribe is reasonably expected to be capable, in the Administrator's judgment, of carrying out the functions to be exercised in a manner consistent with the terms and purposes of this title and of all applicable regulations.

(2) PROVISIONS WHERE TREATMENT AS STATE INAPPROPRIATE—

For any provision of this title where treatment of Indian Tribes as identical to States is inappropriate, administratively infeasible or otherwise inconsistent with the purposes of this title, the Administrator may include in the regulations promulgated under this section, other means for administering such provision in a manner that will achieve the purpose of the provision. Nothing in this section shall be construed to allow Indian Tribes to assume or maintain primary enforcement responsibility for public water systems or for underground injection control in a manner less protective of the health of persons than such responsibility may be assumed or maintained by a State. An Indian tribe shall not be required to exercise criminal enforcement jurisdiction for purposes of complying with the preceding sentence.

APPENDIX 2

Glossary

Absorbed dose—The amount of a chemical that enters the body of an exposed organism.

Absorption—The uptake of water or dissolved chemicals by a cell or an organism.

Absorption factor—The fraction of a chemical making contact with an organism that is absorbed by the organism.

Acceptable daily intake (ADI)—Estimate of the largest amount of chemical to which a person can be exposed on a daily basis that is not anticipated to result in adverse effects (usually expressed in mg/kg/day). (Synonymous with RfD.)

Active transport—An energy-expending mechanism by which a cell moves a chemical across the cell membrane from a point of lower concentration to a point of higher concentration, against the diffusion gradient.

Acute—Occurring over a short period of time; used to describe brief exposures and effects which appear promptly after exposure.

Additive effect—Combined effect of two or more chemicals equal to the sum of their individual effects.

Adsorption—The process by which chemicals are held on the surface of a mineral or soil particle (*compare with* Absorption).

Ambient—Environmental or surrounding conditions.

Animal studies—Investigations using animals as surrogates for humans, on the expectation that results in animals are pertinent to humans.

Antagonism—Interference or inhibition of the effect of one chemical by the action of another chemical.

Assay—A test for a particular chemical or effect.

Bias—An inadequacy in experimental design that leads to results or conclusions not representative of the population under study.

Bioaccumulation—The retention and concentration of a substance by an organism.

Bioassay—Test which determines the effect of a chemical on a living organism.

Bioconcentration—The accumulation of a chemical in tissues of an organism (such as fish) to levels that are greater than the level in the medium (such as water) in which the organism resides (*see* Bioaccumulation).

Biodegradation—Decomposition of a substance into more elementary compounds by the action of microorganisms such as bacteria.

Biotransformation—Conversion of a substance into other compounds by organisms; includes biodegradation.

bw—Body weight.

CAG—Carcinogen Assessment Group.

Cancer—A disease characterized by the rapid and uncontrolled growth of aberrant cells into malignant tumors.

Carcinogen—A chemical which causes or induces cancer.

CAS registration number—A number assigned by the Chemical Abstracts Service to identify a chemical.

Central nervous system—Portion of the nervous system which consists of the brain and spinal cord; CNS.

Chronic—Occurring over a long period of time, either continuously or intermittently; used to describe ongoing exposures and effects that develop only after a long exposure.

Chronic exposure—Long-term, low-level exposure to a toxic chemical.

Clinical studies—Studies of humans suffering from symptoms induced by chemical exposure.

Confounding factors—Variables other than chemical exposure level which can affect the incidence or degree of a parameter being measured.

Cost/benefit analysis—A quantitative evaluation of the costs which would be incurred versus the overall benefits to society of a proposed action such as the establishment of an acceptable dose of a toxic chemical.

Cumulative exposure—The summation of exposures of an organism to a chemical over a period of time.

Degradation—Chemical or biological breakdown of a complex compound into simpler compounds.

Dermal exposure—Contact between a chemical and the skin.

Diffusion—The movement of suspended or dissolved particles from a more concentrated to a less concentrated region as a result of the random movement of individual particles; the process tends to distribute them uniformly throughout the available volume.

Dosage—The quantity of a chemical administered to an organism.

Dose—The actual quantity of a chemical to which an organism is exposed (*see* Absorbed dose).

Dose-response—A quantitative relationship between the dose of a chemical and an effect caused by the chemical.

Dose-response curve—A graphical presentation of the relationship between degree of exposure to a chemical (dose) and observed biological effect or response.

Dose-response evaluation—A component of risk assessment that describes the quantitative relationship between the amount of exposure to a substance and the extent of toxic injury or disease.

Dose-response relationship—The quantitative relationship between the amount of exposure to a substance and the extent of toxic injury produced.

DWEL—Drinking Water Equivalent Level—estimated exposure (in mg/L) which is interpreted to be protective for noncarcinogenic endpoints of toxicity over a lifetime of exposure. DWEL was developed for chemicals that have a significant carcinogenic potential (Group B). Provides risk manager with evaluation on noncancer endpoints, but infers that carcinogenicity should be considered the toxic effect of greatest concern.

Endangerment assessment—A site-specific risk assessment of the actual or potential danger to human health or welfare and the environment from the release of hazardous substances or waste. The endangerment assessment document is prepared in support of enforcement actions under CERCLA or RCRA.

Endpoint—A biological effect used as an index of the effect of a chemical on an organism.

Epidemiologic study—Study of human populations to identify causes of disease. Such studies often compare the health status of a group of persons who have been exposed to a suspect agent with that of a comparable nonexposed group.

Exposure—Contact with a chemical or physical agent.

Exposure assessment—The determination or estimation (qualitative or quantitative) of the magnitude, frequency, duration, route, and extent (number of people) of exposure to a chemical.

Exposure coefficient—Term which combines information on the frequency, mode, and magnitude of contact with contaminated medium to yield a quantitative value of the amount of contaminated medium contacted per day.

Exposure level, chemical—The amount (concentration) of a chemical at the absorptive surfaces of an organism.

Exposure scenario—A set of conditions or assumptions about sources, exposure pathways, concentrations of toxic chemicals and populations (numbers, characteristics and habits) which aid the investigator in evaluating and quantifying exposure in a given situation.

Extrapolation—Estimation of unknown values by extending or projecting from known values.

Gavage—Type of exposure in which a substance is administered to an animal through a stomach tube.

Gram—1/454 of a pound.

Half-life—The length of time required for the mass, concentration, or activity of a chemical or physical agent to be reduced by one-half.

Hazard evaluation—A component of risk assessment that involves gathering and evaluating data on the types of health injury or disease (e.g., cancer) that may be produced by a chemical and on the conditions of exposure under which injury or disease is produced.

Hematopoiesis—The production of blood and blood cells; hemopoiesis.

Hepatic—Pertaining to the liver.

Hepatoma—A malignant tumor occurring in the liver.

High-to-low-dose extrapolation—The process of prediction of low exposure risks to rodents from the measured high exposure-high risk data.

Histology—The study of the structure of cells and tissues; usually involves microscopic examination of tissue slices.

Human equivalent dose—A dose which, when administered to humans, produces an effect equal to that produced by a dose in animals.

Human exposure evaluation—A component of risk assessment that involves describing the nature and size of the population exposed to a substance and the magnitude and duration of their exposure. The evaluation could concern past exposures, current exposures, or anticipated exposures.

Human health risk—The likelihood (or probability) that a given exposure or series of exposures may have or will damage the health of individuals experiencing the exposures.

Incidence of tumors—Percentage of animals with tumors.

Ingestion—Type of exposure through the mouth.

Inhalation—Type of exposure through the lungs.

Integrated exposure assessment—A summation over time, in all media, of the magnitude of exposure to a toxic chemical.

Interspecies extrapolation model—Model used to extrapolate from results observed in laboratory animals to humans.

In vitro studies—Studies of chemical effects conducted in tissues, cells or subcellular extracts from an organism (i.e., not in the living organism).

In vivo studies—Studies of chemical effects conducted in intact living organisms.

Irreversible effect—Effect characterized by the inability of the body to partially or fully repair injury caused by a toxic agent.

Latency—Time from the first exposure to a chemical until the appearance of a toxic effect.

LC_{50}—The concentration of a chemical in air or water which is expected to cause death in 50% of test animals living in that air or water.

LD_{50}—The dose of a chemical taken by mouth or absorbed by the skin which is expected to cause death in 50% of the test animals so treated.

Lesion—A pathological or traumatic discontinuity of tissue or loss of function of a part.

Lethal—Deadly; fatal.

Lifetime exposure—Total amount of exposure to a substance that a human would receive in a lifetime (usually assumed to be 70 years).

Linearized multistage model—Derivation of the multistage model, where the data are assumed to be linear at low doses.

LOAEL—Lowest-observed-adverse-effect level; the lowest dose in an experiment which produced an observable adverse effect.

Malignant—Very dangerous or virulent, causing or likely to cause death.

Margin of safety (MOS)—Maximum amount of exposure producing no measurable effect in animals (or studied humans) divided by the actual amount of human exposure in a population.

Mathematical model—Model used during risk assessment to perform extrapolations.

Metabolism—The sum of the chemical reactions occurring within a cell or a whole organism; includes the energy-releasing breakdown of molecules (catabolism) and the synthesis of new molecules (anabolism).

Metabolite—Any product of metabolism, especially a transformed chemical.

Metastatic—Pertaining to the transfer of disease from one organ or part to another not directly connected with it.

Microgram (μg)—One-millionth of a gram (3.5×10^{-8} oz. = 0.000000035 oz.).

Milligram (mg)—One-thousandth of a gram (3.5×10^{-5} oz. = 0.000035 oz.).

Modeling—Use of mathematical equations to simulate and predict real events and processes.

Monitoring—Measuring concentrations of substances in environmental media or in human or other biological tissues.

Mortality—Death.

MOS. *See* Margin of safety.

MTD—Maximum tolerated dose, the dose that an animal species can tolerate for a major portion of its lifetime without significant impairment or toxic effect other than carcinogenicity.

Multistage model—Mathematical model based on the multistage theory of the carcinogenic process, which yields risk estimates either equal to or less than the one-hit model.

Mutagen—An agent that causes a permanent genetic change in a cell other than that which occurs during normal genetic recombination.

Mutagenicity—The capacity of a chemical or physical agent to cause permanent alteration of the genetic material within living cells.

Necrosis—Death of cells or tissue.

Neoplasm—An abnormal growth or tissue, as a tumor.

Neurotoxicity—Exerting a destructive or poisonous effect on nerve tissue.

NOAEL—No-observed-adverse-effect level; the highest dose in an experiment which did not produce an observable adverse effect.

NOEL—No-observed-effect level; dose level at which no effects are noted.

NTP—National Toxicology Program.

Oncology—Study of cancer.

One-hit model—Mathematical model based on the biological theory that a single "hit" of some minimum critical amount of a carcinogen at a cellular target—namely, DNA—can initiate an irreversible series of events, eventually leading to a tumor.

Oral—Of the mouth; through or by the mouth.

Pathogen—Any disease-causing agent, usually applied to living agents.

Pathology—The study of disease.

Permissible dose—The dose of a chemical that may be received by an individual without the expectation of a significantly harmful result.

Pharmacokinetics—The dynamic behavior of chemicals inside biological systems; it includes the processes of uptake, distribution, metabolism, and excretion.

Population at risk—A population subgroup that is more likely to be exposed to a chemical, or is more sensitive to a chemical, than is the general population.

Potency—Amount of material necessary to produce a given level of a deleterious effect.

Potentiation—The effect of one chemical to increase the effect of another chemical.

ppb—Parts per billion.

ppm—Parts per million.

Prevalence study—An epidemiological study which examines the relationships between diseases and exposures as they exist in a defined population at a particular point in time.

Prospective study—An epidemiological study which examines the development of disease in a group of persons determined to be presently free of the disease.

Qualitative—Descriptive of kind, type or direction, as opposed to size, magnitude or degree.

Quantitative—Descriptive of size, magnitude or degree.

Receptor—(1) In biochemistry: a specialized molecule in a cell that binds a specific chemical with high specificity and high affinity. (2) In exposure assessment: an organism that receives, may receive, or has received environmental exposure to a chemical.

Renal—Pertaining to the kidney.

Reservoir—A tissue in an organism or a place in the environment where a chemical accumulates, from which it may be released at a later time.

Retrospective study—An epidemiological study which compares diseased persons with nondiseased persons and works back in time to determine exposures.

Reversible effect—An effect which is not permanent, especially adverse effects which diminish when exposure to a toxic chemical is ceased.

RfD—Reference dose; the daily exposure level which, during an entire lifetime of a human, appears to be without appreciable risk on the basis of all facts known at the time. (Synonymous with ADI.)

Risk—The potential for realization of unwanted adverse consequences or events.

Risk assessment—A qualitative or quantitative evaluation of the environmental and/or health risk resulting from exposure to a chemical or physical agent (pollutant); combines exposure assessment results with toxicity assessment results to estimate risk.

Risk characterization—Final component of risk assessment that involves integration of the data and analysis involved in hazard evaluation, dose-response evaluation, and human exposure evaluation to determine the likelihood that humans will experience any of the various forms of toxicity associated with a substance.

Risk estimate—A description of the probability that organisms exposed to a specified dose of chemical will develop an adverse response (e.g., cancer).

Risk factor—Characteristic (e.g., race, sex, age, obesity) or variable (e.g., smoking, occupational exposure level) associated with increased probability of a toxic effect.

Risk management—Decisions about whether an assessed risk is sufficiently high to present a public health concern and about the appropriate means for control of a risk judged to be significant.

Risk specific dose—The dose associated with a specified risk level.

Route of exposure—The avenue by which a chemical comes into contact with an organism (e.g., inhalation, ingestion, dermal contact, injection).

Safe—Condition of exposure under which there is a "practical certainty" that no harm will result in exposed individuals.

Sink—A place in the environment where a compound or material collects (*see* Reservoir).

Sorption—A surface phenomenon which may be either absorption or adsorption, or a combination of the two; often used when the specific mechanism is not known.

Stochastic—Based on the assumption that the actions of a chemical substance results from probabilistic events.

Stratification—(1) The division of a population into subpopulations for sampling purposes. (2) The separation of environmental media into layers, as in lakes.

Subchronic—Of intermediate duration, usually used to describe studies or levels of exposure between 5 and 90 days.

Synergism—An interaction of two or more chemicals which results in an effect that is greater than the sum of their effects taken independently.

Systemic—Relating to whole body, rather than its individual parts.

Systemic effects—Effects observed at sites distant from the entry point of a chemical due to its absorption and distribution into the body.

Teratogenesis—The induction of structural or functional development abnormalities by exogenous factors acting during gestation; interference with normal embryonic development.

Teratogenicity—The capacity of a physical or chemical agent to cause nonhereditary congenital malformations (birth defects) in offspring.

Therapeutic index—The ratio of the dose required to produce toxic or lethal effect to dose required to produce nonadverse or therapeutic response.

Threshold—The lowest dose of a chemical at which a specified measurable effect is observed and below which it is not observed.

Time-weighted average—The average value of a parameter (e.g., concentration of a chemical in air) that varies over time.

Tissue—A group of similar cells.

Toxicant—A harmful substance or agent that may injure an exposed organism.

Toxicity—The quality or degree of being poisonous or harmful to plant, animal or human life.

Toxicity assessment—Characterization of the toxicological properties and effects of a chemical, including all aspects of its absorption, metabolism, excretion and mechanism of action, with special emphasis on establishment of dose-response characteristics.

Transformation—Acquisition by a cell of the property of uncontrolled growth.

Tumor incidence—Fraction of animals having a tumor of a certain type.

Uncertainty factor—A number (equal to or greater than one) used to divide NOAEL or LOAEL values derived from measurements in animals or small groups of humans, in order to estimate a NOAEL value for the whole human population.

Unit cancer risk—Estimate of the lifetime risk caused by each unit of exposure in the low exposure region.

Upper-bound estimate—Estimate not likely to be lower than the true risk.

Volatile—Readily vaporizable at a relatively low temperature.

List of Authors

Edward J. Calabrese, Director, Northeast Regional Environmental Public Health Center, Division of Public Health, University of Massachusetts, Amherst, Massachusetts 01003

Joseph A. Cotruvo, Director, Criteria and Standards Division, Office of Drinking Water, U.S. Environmental Protection Agency, 401 M Street SW, Washington, DC 20460

Charles E. Gilbert, Assistant Director, Northeast Regional Environmental Public Health Center, Division of Public Health, University of Massachusetts, Amherst, Massachusetts 01003

Peter C. Karalekas, Jr., Chief, Water Supply Section, U.S. Environmental Protection Agency, JFK Federal Building (WSB-2113), Boston, Massachusetts 02203

Charles D. Larson, Chief, Technical Assistance Section, Water Supply Branch, U.S. Environmental Protection Agency, JFK Federal Building (WST-2113), Boston, Massachusetts 02203

Edward V. Ohanian, Chief, Health Effects Branch, Criteria and Standards Division, Office of Drinking Water, U.S. Environmental Protection Agency, 401 M Street SW (WH-550), Washington, DC 20460

Marlene Regelski, Environmental Protection Specialist, Office of Water, U.S. Environmental Protection Agency, 401 M Street SW (WH-595), Washington, DC 20460

John R. Trax, Chief, Water Supply Branch, U.S. Environmental Protection Agency, 401 M Street SW (WH-550), Washington, DC 20460

Bailus Walker, Jr., Professor of Environmental Health and Toxicology, School of Public Health, State University of New York, Empire State Plaza, 2523 Corning Tower, Albany, NY 12237

Index